T0076049

METHODS IN MOLECULAR BIOLOGY™

Series Editor
John M. Walker
School of Life Sciences
University of Hertfordshire
Hatfield, Hertfordshire, AL10 9AB, UK

For further volumes:
http://www.springer.com/series/7651

Gene Expression Profiling

Methods and Protocols

Second Edition

Edited by

Lorraine O'Driscoll

School of Pharmacy and Pharmaceutical Sciences, Trinity College Dublin, Dublin, Ireland

Editor
Lorraine O'Driscoll
School of Pharmacy and Pharmaceutical Sciences
Trinity College Dublin
Dublin, Ireland
lodrisc@tcd.ie

ISSN 1064-3745 e-ISSN 1940-6029
ISBN 978-1-61779-288-5 e-ISBN 978-1-61779-289-2
DOI 10.1007/978-1-61779-289-2
Springer New York Dordrecht Heidelberg London

Library of Congress Control Number: 2011935371

Printed on acid-free paper

Humana Press is part of Springer Science+Business Media (www.springer.com)

Preface

It was with great pleasure that I accepted the invitation to be Editor for this volume of *Methods in Molecular Biology*. This book collates chapters by experts on a wide range of topics relevant to gene expression profiling.

Understanding gene expression and how it changes under normal and pathological conditions is essential to our understanding of the fundamentals of cell biology through to the targeted treatment of disease. This book compiles protocols, written by experts in their respective fields, for a broad range of techniques, currently available and being further developed, for the analysis of gene expression at the DNA, RNA, and protein levels.

In summary, the topics addressed in this volume of *Methods in Molecular Biology* span the field of gene expression from basic to advanced methodologies, including step-by-step protocols which the reader can easily follow, as well as trouble-shooting tips and tricks to help ensure the success of their application. The chapter by Gurvich and Skoblov brings us through the fundamentals of polymerase chain reaction (PCR), extending to multiplex approaches. Rani and O'Driscoll advance on this strong basis to the application of PCR for the analysis of specimens where very limited amounts of starting material exist, i.e., in the extracellular environment. Considering the importance of being able to globally co-analyze all mRNAs transcribed, Mehta outlines microarray methods. Following microarray analysis, researchers often find progression to the stage of being able to correctly analyze their data to be a substantial challenge. For this reason, Mehta and Rani proceed to explain software and tools for the analysis of such datasets. Friel, Crown, and O'Driscoll detail gene expression analysis specifically in relation to cancer cells and circulating tumor cells. As formalin-fixed paraffin-embedded tissue is quite extensively available for gene expression studies, relative to the availability of fresh/frozen tissue, April and Fan have described methods for such studies, using the whole genome DASL assay. Moving into the world of microRNAs (miRNAs), Hennessy and O'Driscoll outline methods for their determination, including procedures which are applicable to the assessment of both intracellular and extracellular miRNAs. The final steps in our journey through profiling of gene expression involve analysis at the protein level. Here, initially considering extracted proteins, Germano and O'Driscoll describe basic, but essential, methods for assessing proteins by Western blotting, while Meleady advances this to more global protein studies that involve 2D gel electrophoresis and mass spectrometry technologies. A powerful advancement for in situ protein evaluation is our ability to construct "user-friendly" arrays representative of cell lines and/or tissues of interest. Experts in this field, Gately, Kerr, and O'Byrne, detail their optimized methods for design, construction, and analysis of such cell/tissue arrays. Basic and advanced immunohistochemistry and immunofluorescence techniques are clearly described by Katikreddy and O'Sullivan, while Hanrahan, Harris, and Egan detail laser scanning confocal microscopy in such a way as to make it less daunting to consider, when analysis and co-analysis of proteins in situ is desired. Rani, O'Brien, Kelleher, Corcoran, Germano, Radomski, Crown, and O'Driscoll describe procedures for the isolation of exosomes and their subsequent gene expression analysis as potential representative of the cells from which they have been secreted.

Finally, Buckley, Davies, and Ehrhardt take us through step-by-step approaches to the successful application of atomic force microscopy and high-content analysis for analyzing expression of particular genes of interest.

I hope that this collection of clearly described and illustrated chapters will be helpful to researchers in academia, in hospitals, and in industry who are interested in applying techniques, whether they be very basic or very advanced, for the analysis of gene expression. I also hope this will be of use to those who want to review progress in this very exciting, evolving field. Mostly I hope it will contribute, in some way, to a better understanding of gene expression changes that occur under normal and pathological conditions which can advance progress toward better treatments for those suffering from disease.

Dublin, Ireland ***Lorraine O'Driscoll***

Contents

Preface v
Contributors ix

1 Real-Time PCR and Multiplex Approaches 1
Olga L. Gurvich and Mikhail Skoblov

2 Reverse-Transcriptase Polymerase Chain Reaction to Detect Extracellular mRNAs 15
Sweta Rani and Lorraine O'Driscoll

3 Microarray Analysis of mRNAs: Experimental Design and Data Analysis Fundamentals 27
Jai Prakash Mehta

4 Software and Tools for Microarray Data Analysis 41
Jai Prakash Mehta and Sweta Rani

5 Analysis of Gene Expression as Relevant to Cancer Cells and Circulating Tumour Cells 55
Anne M. Friel, John Crown, and Lorraine O'Driscoll

6 Gene Expression Profiling in Formalin-Fixed, Paraffin-Embedded Tissues Using the Whole-Genome DASL Assay 77
Craig S. April and Jian-Bing Fan

7 MicroRNA Expression Analysis: Techniques Suitable for Studies of Intercellular and Extracellular MicroRNAs 99
Erica Hennessy and Lorraine O'Driscoll

8 Western Blotting Analysis as a Tool to Study Receptor Tyrosine Kinases 109
Serena Germano and Lorraine O'Driscoll

9 2D Gel Electrophoresis and Mass Spectrometry Identification and Analysis of Proteins 123
Paula Meleady

10 Design, Construction, and Analysis of Cell Line Arrays and Tissue Microarrays for Gene Expression Analysis 139
Kathy Gately, Keith Kerr, and Ken O'Byrne

11 Immunohistochemical and Immunofluorescence Procedures for Protein Analysis 155
Kishore Reddy Katikireddy and Finbarr O'Sullivan

12 Advanced Microscopy: Laser Scanning Confocal Microscopy 169
Orla Hanrahan, James Harris, and Chris Egan

13 Isolation of Exosomes for Subsequent mRNA, MicroRNA, and Protein Profiling 181
Sweta Rani, Keith O'Brien, Fergal C. Kelleher, Claire Corcoran, Serena Germano, Marek W. Radomski, John Crown, and Lorraine O'Driscoll

14 Atomic Force Microscopy and High-Content Analysis: Two Innovative Technologies for Dissecting the Relationship Between Epithelial–Mesenchymal Transition-Related Morphological and Structural Alterations and Cell Mechanical Properties 197
Stephen T. Buckley, Anthony M. Davies, and Carsten Ehrhardt

Index *209*

Contributors

CRAIG S. APRIL • *Illumina, Inc., San Diego, CA, USA*
STEPHEN T. BUCKLEY • *School of Pharmacy & Pharmaceutical Sciences, Trinity College Dublin, Dublin, Ireland*
CLAIRE CORCORAN • *School of Pharmacy & Pharmaceutical Sciences & MTCI, Trinity College Dublin, Dublin, Ireland*
JOHN CROWN • *Molecular Therapeutics for Cancer Ireland (MTCI), Dublin City University, Dublin, Ireland*
ANTHONY M. DAVIES • *Department of Clinical Medicine, Trinity College Dublin, Dublin, Ireland*
CHRIS EGAN • *Smurfit Institute of Genetics, Trinity College Dublin, Dublin, Ireland*
CARSTEN EHRHARDT • *School of Pharmacy & Pharmaceutical Sciences, Trinity College Dublin, Dublin, Ireland*
JIAN-BING FAN • *Illumina, Inc., San Diego, CA, USA*
ANNE M. FRIEL • *School of Pharmacy and Pharmaceutical Sciences & MTCI, Trinity College Dublin, Dublin, Ireland*
KATHY GATELY • *Department of Clinical Medicine, Thoracic Oncology Research Group, Institute of Molecular Medicine, Trinity Centre for Health Sciences, Dublin, Ireland*
SERENA GERMANO • *School of Pharmacy and Pharmaceutical Sciences, Trinity College Dublin, Dublin, Ireland*
OLGA L. GURVICH • *Moscow State University, Moscow, Russia*
ORLA HANRAHAN • *School of Biochemistry & Immunology, Trinity College Dublin, Dublin, Ireland*
JAMES HARRIS • *School of Biochemistry & Immunology, Trinity College Dublin, Dublin, Ireland*
ERICA HENNESSY • *National Institute for Cellular Biotechnology, Dublin City University, Dublin, Ireland*
KISHORE REDDY KATIKIREDDY • *National Institute of Cellular Biotechnology, Dublin City University, Dublin, Ireland*
KEITH KERR • *Department of Pathology, Aberdeen University Medical School & Aberdeen Royal Infirmary, Foresterhill, Aberdeen, UK*
FERGAL C. KELLEHER • *Department of Medical Oncology, St. Vincent's University Hospital, Dublin, Aberdeen, UK*
JAI PRAKASH MEHTA • *Conway Institute, University College Dublin, Dublin, Ireland*
PAULA MELEADY • *National Institute for Cellular Biotechnology, Dublin City University, Dublin, Ireland*
KEITH O'BRIEN • *School of Pharmacy & Pharmaceutical Sciences, Trinity College Dublin, Dublin, Ireland*
KEN O'BYRNE • *Department of Clinical Medicine, Thoracic Oncology Research Group, Institute of Molecular Medicine, Trinity Centre for Health Sciences, Dublin, Ireland*

LORRAINE O'DRISCOLL • *School of Pharmacy and Pharmaceutical Sciences, Trinity College Dublin, Dublin, Ireland*
FINBARR O'SULLIVAN • *National Institute of Cellular Biotechnology, Dublin City University, Dublin, Ireland*
MAREK W. RADOMSKI • *School of Pharmacy & Pharmaceutical Sciences, Trinity College Dublin, Dublin, Ireland*
SWETA RANI • *School of Pharmacy & Pharmaceutical Sciences, Panoz Institute, Trinity College Dublin, Dublin, Ireland*
MIKHAIL SKOBLOV • *Research Centre for Medical Genetics, Russian Academy of Medical Sciences, Moscow, Russia*

Chapter 1

Real-Time PCR and Multiplex Approaches

Olga L. Gurvich and Mikhail Skoblov

Abstract

Analysis of RNA expression levels by real-time reverse-transcription (RT) PCR has become a routine technique in diagnostic and research laboratories. Monitoring of DNA amplification can be done using fluorescent sequence-specific probes, which generate signal only upon binding to their target. Numerous fluorescent dyes with unique emission spectra are available and can be used to differentially label probes for various genes. Such probes can be added to the same PCR amplification reaction for simultaneous detection of multiple targets in a single assay. Such multiplexing is advantageous, since it markedly increases throughput and decreases costs and labor. Here, we describe application of multiplex real-time RT-PCR using TaqMan probes in the analysis of relative expression levels of a novel tumor-associated gene CUG2 in cell lines and tissue samples.

Key words: Multiplex, Real-time PCR, Gene expression, TaqMan probes, CUG2, C6orf173

1. Introduction

The real-time, reverse-transcription (RT) quantitative (q) PCR is the established method of choice for reliable, rapid, and sensitive measurement of gene expression (1). It is widely used in nucleic acid research and increasingly so in clinical settings. Its ability to accurately detect levels of endogenous mRNAs, miRNAs (2) as well as RNA from viral pathogens makes it a unique tool for early diagnosis of human disease, such as cancers (3–6) and viral infections (7, 8). Moreover, it can be used to screen for biomarkers predicative of beneficial therapeutic interventions (9) or graft rejection and immunosuppresion in transplant biology (10).

Original use of intercalating dyes, such as SYBR green, to monitor DNA amplification, enabled measurement of a single mRNA in a single tube. Thus, to reliably quantify expression of a

Lorraine O'Driscoll (ed.), *Gene Expression Profiling: Methods and Protocols*, Methods in Molecular Biology, vol. 784,
DOI 10.1007/978-1-61779-289-2_1, © Springer Science+Business Media, LLC 2011

single gene, it was necessary to set up multiple reactions to develop standard curves for the reference gene and the gene of interest. Another major issue with SYBR green detection is that all amplified products intercalate the dye and increase the generated signal. As a result, sensitivity of the technique made any pipetting errors as well as nonspecific amplification detrimental to the accurate estimation of expression levels. Advances in fluorescent chemistries (11) and the use of sequence-specific probes made it possible to determine the expression of a reference gene and the gene(s) of interest in a single tube, thus requiring less sample handling and providing internal normalization. Probes used in multiplex assays are conjugated with fluorescent reporter dyes, which vary in their absorption and emission spectra. The number of genes that can be analyzed in the same tube is limited by spectral properties of the available fluorophores, which should have nonoverlapping emission spectra for separate detection. The majority of modern spectrofluorometric thermal cyclers allow the measurement of fluorescence in four different wavelength channels (although six-channel machines do exist), thereby enabling the concurrent detection of up to four different DNA targets. A choice of dyes optimized for detection in each channel (Table 1) is available for distinctive labeling of specific probes.

The most widely used probes to monitor DNA amplification in real-time PCR in clinical settings are TaqMan or hydrolysis probes (12). These probes are labeled at one end with a reporter fluorescent dye and on the other with a fluorescence quencher, which must exhibit spectral overlap with the fluorophore (Table 1). Quencher absorbs energy emitted by the flourophore through

Table 1
Most commonly used fluorescent dyes and their optimal quenchers for TaqMan probes in multiplex PCR

Channel detection wavelength (nm)	Dyes used for channel [a]	Corresponding quencher [b]
520	*FAM, SybrGreen* [c]	*Dabcyl, BHQ1*
550	*HEX, SIMA, VIC, JOE*	*BHQ1, BHQ2*
580	TAMRA	BHQ2
610	*ROX, Cy3.5*	*BHQ2*
670	*Cy5*	*BHQ2, BHQ3*
705	Cy5.5	BHQ3

[a] Where multiple fluorophores are listed for specific channel, either of them can be used in multiplex detection
[b] Same quencher can be used with two different fluorophores in multiplex PCR
[c] Four main detection channels are indicated in *italics*

fluorescent resonance energy transfer (FRET); the quenchers, currently used in multiplex PCR, reemit energy as heat (dark quencher). When the probe is intact, the proximity of the quencher to the reporter dye inhibits the fluorescence signal. The T_m of the probe is usually 8–10°C higher than that of the primers, which allows the probe to anneal prior to extension. During DNA synthesis, the 5′-exonuclease activity of the Taq polymerase hydrolyses the bound probe and releases the dye from the quencher, relieving quenching effect. The level of detected fluorescence is therefore proportional to the amounts of newly synthesized DNA. A number of other types of probes are being used in real-time PCR as well, including hybridization probes and molecular beacons (13). Hybridization probes consist of two single-dye labeled oligonucleotides which bind to adjacent targets in the amplified region, thereby bringing the two dyes in close proximity and inducing FRET. Molecular beacon probes are designed as hairpins, with target sequence located in the loop and fluorescent dye and the quencher conjugated to the 5′ and 3′ of the oligonucleotide. In the intact, hairpin state, fluorescence is quenched; binding of the probe to the amplified sequence forces fluorophore and quencher apart, and releases fluorescent signal. Another widely used probes are Scorpion primers, which combine PCR primer and specific probe in one sequence (14). The structure of Scorpion primers promotes unimolecular probing mechanism; consequently, they can perform better than TaqMan probes or molecular beacons particularly under fast cycling conditions (15). Described probes can be used not only to quantify levels of DNA, but to discriminate between alleles and detect mutations as well.

Here, TaqMan probes are used in multiplex real-time RT-PCR to analyze expression levels of a novel tumor-associated gene CUG2 (cancer-upregulated gene 2) in HEK293T cells and tissue samples. CUG2, or C6orf173 (chromosome 6 open reading frame 173) was recently identified as an expressed sequence tag (EST) that exhibits significant differential expression in multiple human cancer types (16). CUG2 overexpression resulted in oncogenic transformation of cell lines and produced tumors in nude mice. It has recently been shown to colocalize with human centromeric markers and its inhibition induced aberrant cell division, suggesting a role in chromosome segregation during mitosis (17). The HPRT1 gene is used as a reference gene. The HPRT1 gene encodes an enzyme, hypoxanthine phosphoribosyltransferase, which plays a central role in the generation of purine nucleotides through the purine salvage pathway. Several studies have indicated that HPRT1 mRNA is stably expressed across a variety of tissues and tumor samples, and it has been recommended for standardization of real-time RT-PCR data (18–20).

2. Materials

2.1. Cell Culture, RNA Purification, and Quality Control

1. Growth medium: Dulbecco's modified Eagle's medium (DMEM; Paneco, Russia) supplemented with 10% fetal bovine serum (Paneco, Russia).
2. Solution of trypsin (0.25%) and ethylenediamine tetraacetic acid (EDTA; 1 mM) (Paneco, Russia).
3. Versene buffer: 1.47 mM KH_2PO_4, 4.29 mM Na_2HPO_4, 137 mM NaCl, 2.68 mM KCl, and 2 mM EDTA.
4. RNeasy Protect Mini Kit (Qiagen).
5. Biophotometer (Eppendorf).
6. Agarose gel electrophoresis accessories:
   (a) peqGOLD Universal-Agarose (PeqLab).
   (b) TAE buffer: 40 mM Tris acetate, 1 mM EDTA. To make 50× stock dissolve 242 g Tris base in approximately 750 ml deionized (d) H_2O, add 57.1 ml glacial acetic acid, 100 ml of 0.5 M EDTA (pH 8.0), and adjust with dH_2O to a final volume of 1 L. Store at room temperature. The pH is not adjusted and should be about 8.5. Dilute to 1× with dH_2O prior to use.
   (c) Ethidium bromide 10 mg/ml (Sigma).
   (d) 10× loading buffer: 50% glycerol, 0.25% bromophenol blue, and 0.25% xylene cyanole in 1× TAE buffer.
   (e) 1 kb DNA Ladder (Promega).

2.2. cDNA Synthesis

1. ImProm-II Reverse Transcription System (Promega).
2. cDNA Synthesis Primer (CDS Primer; 10 mM). 5′-$(T)_{25}$ V N-3′ (N = A, C, G, or T; V = A, G, or C) synthesized by DNA-Synthesis (Moscow, Russia).

2.3. Primer/Probes Design

Primers and probes for the HPRT1 and CUG2 were designed using Oligo.4 software (www.oligo.net), synthesized by DNA-Synthesis (Moscow, Russia), and stored as 50 mM stocks diluted in nuclease-free H_2O at −20°C. Probes should be kept in dark. Sequences of primers and probes are provided in Table 2.

2.4. PCR Reagents

1. Taq DNA polymerase with 5′→3′ exonuclease activity (Dialat, Ltd, Russia), 2 mM dNTP, 25 mM $MgCl_2$, and 10× buffer for Taq Pol (700 mM Tris–HCl, pH 8.6, 166 mM $(NH_4)_2SO_4$).
2. Real-Time PCR Thermal cycler, RotorGene 3000 (Corbett Research) equipped with four channels: FAM/SYBR (Excite – 470 nm, detect – 510 nm), JOE (Excite – 530 nm, detect – 555 nm), ROX (Excite – 585 nm, detect – 610 nm), and Cy5 (Excite – 625 nm, detect – 660 nm).

Table 2
Sequences of primers and probes for real-time PCR

Gene name and accession number	Primer and probe sequences	T_m (°C)
HPRT1 (NM_000194.2)	Forward: 5′-CTGTGGCCATCTGCTTAGTA-3′	60
	Reverse: 5′-TAGTGCTGTGGTTTAAGAGAAT-3′	60
	Probe: 5′-FAM-TAGATCCATTCCTATGACTGTAGAT-BHQ1-3′	68
C6orf173 (NM_001012507)	Forward: 5′-CGGAGCTTGTGTGCGATACA-3′	62
	Reverse: 5′-CAGTTCAGATGGACCAATAAGT-3′	62
	Probe: 5′-ROX-GAGGCAGCTGGTACTTGACAGAGA-BHQ2-3′	74

3. Methods

3.1. Sample Preparation, RNA Purification, and Quality Control

3.1.1. RNA Isolation from Cell Cultures

1. One milliliter aliquots of HEK 293T cells are routinely stored in cryovials at 10^6 cells/ml in growth medium with 10% DMSO in the liquid nitrogen or at −80°C for short-term storage.
2. Cells are removed from storage and immediately thawed in the 37°C water bath. If cells cannot be transferred directly from the storage to the water bath, they should be kept on dry ice while in transit.
3. Once thawed, cells are gently resuspended in 10 ml growth medium.
4. Cells are centrifuged at ~200 × *g* for 5 min and supernatant is discarded.
5. Cells are resuspended in complete growth medium, counted, and plated at 3×10^5 cells/ml.
6. Cells are grown on 75-mm² plates till confluency for 2–3 days. Growth medium is discarded and cells washed with 10 ml of Versene buffer.
7. Six hundred microliters of Buffer RLT from RNeasy Protect Mini Kit is added directly to the cells on the plate. The buffer is swirled around the plate so that all cells are covered. After 5 min of incubation, cells are scraped off the plate and the solution is transferred to an Eppendorf tube.
8. Total RNA is then extracted according to the Qiagen protocol for the RNeasy Kit (see Note 1 for general RNA handling techniques).

3.1.2. RNA Isolation from Tissues

1. Tissue samples are generally stored at –80°C. When removed from the freezer, tissue should be kept as cold as possible, preferably on dry ice or liquid nitrogen (see Note 2 for precautions during tissue handling).
2. Tissue is sectioned using cryostat or sharp blade. Sections are collected into an Eppendorf tube and covered with 600 μl of Buffer RLT from the RNeasy Protect Mini Kit.
3. Total RNA is then extracted according to the Qiagen protocol for the RNeasy Kit (see Note 3 for an alternative method).

3.1.3. Analysis of RNA Quality and Quantity

1. At the final step of purification using the Qiagen RNeasy Kit, RNA is eluted from the purification column with 50 μl of RNase-free water.
2. One microliter of eluted RNA is used to analyze RNA concentration and the A260/A280 ratio on a spectrophotometer. Generally, RNA yield from HEK293T cells grown to confluency on 75-mm^2 plates is about 80 μg. A ratio of A260/A280 between 1.8 and 2.1 indicates high purity of extracted RNA.
3. RNA quality is further analyzed by electrophoresis in a 1% agarose gel (see also Notes 4 and 5). To prepare gel, 1 g of agarose is added to 100 ml of 1×TAE and boiled in the microwave oven until agarose particles are dissolved. Solution is cooled to 60°C, mixed with 5 μl of ethidium bromide, and poured into the gel tray with the appropriate comb. Once solidified, the gel is placed into the electrophoresis chamber, and 1× TAE buffer added so that the gel is covered. Samples and 5 μl of 1 kb DNA Ladder are mixed with 10× loading dye, loaded on the gel, and electrophoresed for approximately 45 min at 100 V. The gel is exposed to UV irradiation and photographed. Intact total RNA run on a denaturing gel will have sharp, clear 28S and 18S rRNA bands (Fig. 1). The 28S rRNA band should be approximately twice as intense as the 18S rRNA, and is a good indication that the RNA is completely intact. Partially degraded RNA will have a smeared appearance, will lack the sharp rRNA bands, or will not exhibit the 2:1 ratio of high-quality RNA. Completely degraded RNA will appear as a very low molecular weight smear.

3.2. cDNA Synthesis

1. On ice, mix 1 μg of isolated RNA with 10 pmol cDNA Synthesis Primer in 5 μl of RNase-free H_2O (see Note 6).
2. The primer/template mix is denatured for 10 min at 70°C and then chilled on ice.
3. Meanwhile, the reverse-transcription reaction mix is prepared: 4 μl ImProm-II 5× reaction buffer, 3 μl 25 mM $MgCl_2$, 1 μl of dNTP Mix (final concentration 0.5 mM each dNTP), 20 U of RNasin ribonuclease inhibitor, 1 μl ImProm-II reverse transcriptase – 1.0 μl, and RNase-free H_2O to 15 μl. The assembled reaction is kept on ice till the next step.

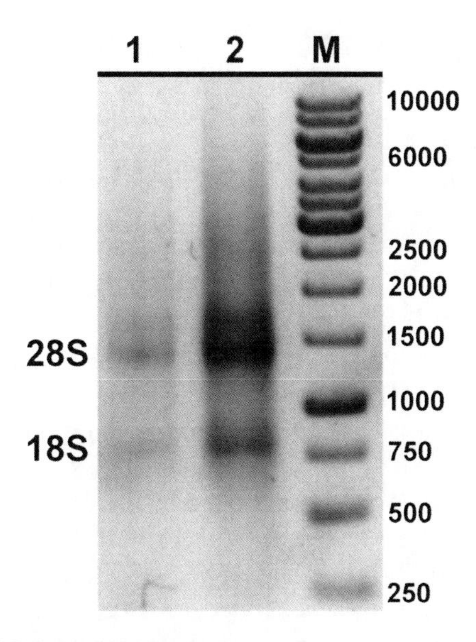

Fig. 1. Analysis of RNA integrity by agarose gel electrophoresis. The same RNA sample was loaded into *Lane 1* (200 ng) and *Lane 2* (1 μg).

4. Add reverse-transcription reaction mix to the primer/template mix; place reaction at 25°C (usually room temperature is appropriate), and incubate for 5 min.
5. The reaction is incubated at 42°C for 1 h and quenched at 70°C for 15 min.
6. The obtained cDNA is diluted to 200 μl with H_2O.

3.3. Design of Primers and Probes

Prior to primer selection, the uniqueness of the target mRNA sequence was analyzed using BLAST to check for potential pseudogenes or repeated sequences in the genome. Primers and probes were selected using Oligo.4 so that the primers' T_m are between 58 and 64°C, while T_m for the probes are 66–78°C. Selected primers and probes have minimal potential to form primer dimers and stem loop structures and low potential for cross-hybridization. Primers should not contain simple repeats; nor are they complementary to repetitive sequences in the genome. Partial complementarity with exact matches at the 3′ end that can cause false priming was avoided. The reverse primer is selected within 300 nucleotides from the forward primer (192 nts for HPRT1 and 204 nts for CUG2). For multiplex detection, probes are labeled with dyes that can be detected in different channels and appropriate quencher (see Table 1). Thus, the designed probe for CUG2 was labeled with carboxy-X-rhodamine (ROX) and Black Hole Quencher 2 (BHQ2) and for HRTP1 with 5-carboxyfluorescein (FAM) and Black Hole Quencher 1 (BHQ1) (Table 2).

3.4. Real-Time PCR

1. PCR reactions are assembled (see Note 7 for general considerations in PCR set up) in triplicates on ice in PCR tubes in a total volume of 25 μl each containing: 2.5 μl 10× Taq buffer, 2.5 μl dNTP (2 mM), 2.5 μl $MgCl_2$ (25 mM), 10 μl cDNA, 2 U of Taq DNA polymerase, and 10 pmol of each HPRT1Forward, HPRT1Reverse, HPRT1probe, CUG2Forward, CUG2Reverse, and CUG2probe. Generally, the Master Mix is prepared (see Note 8); it has all components (except for the DNA template) in amounts sufficient for the total number of reactions (i.e., three multiplied by the number of tested and control samples plus one extra to account for any pipetting inaccuracies). Then, the Master Mix is split for the number of tested samples and cDNA is added in trice the amount; these are subsequently split into three triplicate reactions.
2. The reaction in which cDNA is substituted with an equal amount of H_2O is also assembled and serves as a negative control. This control enables detection of any possible contamination with the PCR product. Another negative control, to assure complete absence of gDNA contamination involves regular PCR using purified RNA as a template.
3. Tubes are transferred directly from ice to the real-time PCR thermal cycler. Amplification is carried out for 45 cycles with denaturation step at 94°C for 10 s, annealing step at 60°C for 5 s, and extension step at 72°C for 10 s (see Note 9).
4. Analysis of obtained data is performed using software "Rotor-Gene 6" supplied with the real-time PCR thermal cycler (see Note 10). If the amplification curve for one of the triplicates substantially differs from the other two, it is omitted from the analysis. Based on the amplification efficiency, the program calculates the threshold cycle, C(T), for each gene and then calculates the difference in threshold cycles, ΔC(T), between the test and the reference gene. The relative difference in expression of the test and reference gene equals $2\Delta^{C(T)}$. Thus, for HEK293T cells ΔC(T) for CUG2 and HPRT1 equals 0.75 (Fig. 2a). This means that amplification efficiency of HPRT1 is $2^{0.75} = 1.68$ times higher than that of CUG2 in HEK293T cells. Expression of the house keeping gene, HPRT1, is assumed to be constant across different samples and equals 100%. Expression of CUG2 is calculated relative to HPRT1, 100%/1.68 = 59.5% in HEK293T. Similar analysis was performed from amplification curves obtained with cDNA from tissues, like myoma and heart (Fig. 2b, c) to compare relative expression of CUG2 in different tissue samples (Fig. 2d) (see also Note 11).

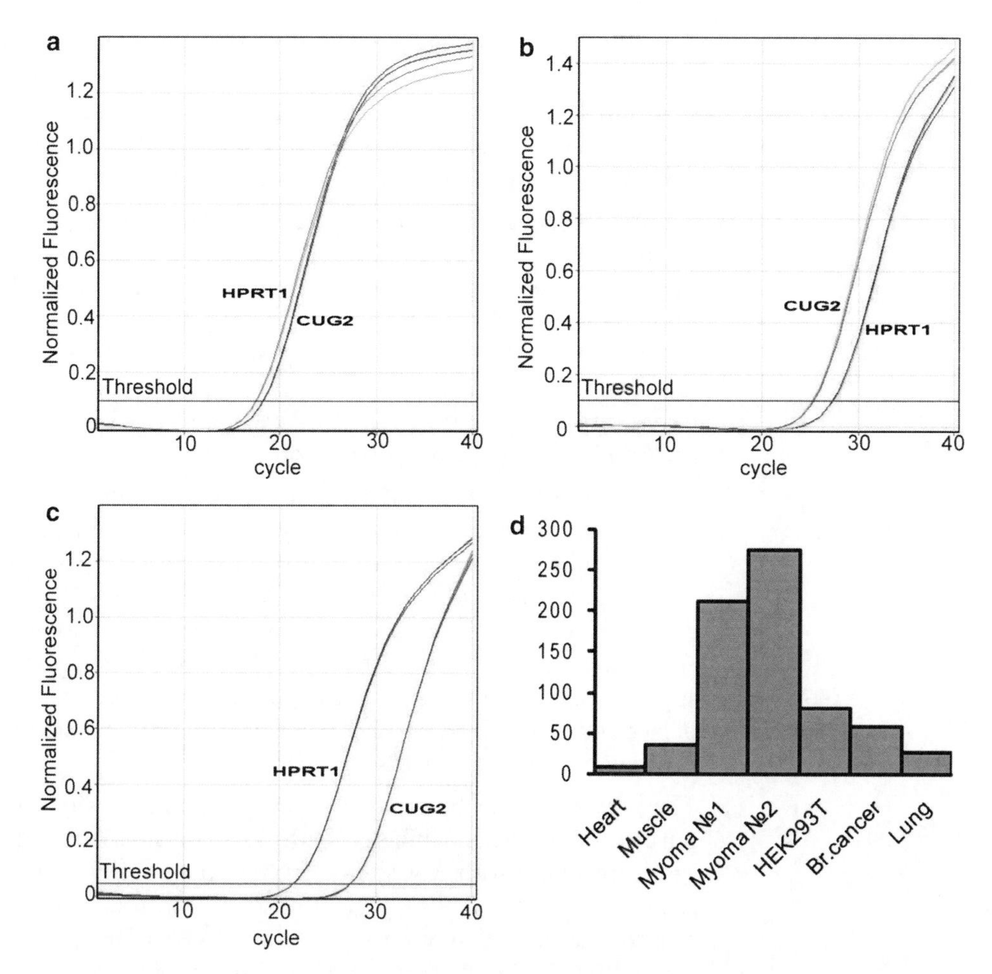

Fig. 2. Quantitative real-time PCR analysis of CUG2 expression in HEK293T cells and tissue samples. Amplification curves for CUG and HPRT1 transcripts from mRNA isolated from (**a**) HEK293T cells, (**b**) myoma, and (**c**) heart. (**d**) Quantitation of relative CUG2 abundance in HEK293T cells, normal (heart, muscle, and lung) and tumor (myoma and breast cancer) tissues. Expression of reference gene HPRT1 in all tissues equals 100%.

4. Notes

1. Good RNA handling techniques are essential. These include wearing clean gloves, using RNase-free equipment and reagents, and filter-tips. Area and equipment should be regularly decontaminated with RNaseZap solution (Ambion) or equivalent. All H_2O used throughout the protocol was filtered through the 0.2 μm filter (Millipore) and treated with DEPC (Sigma), if used for RNA.
2. The RNA isolation step is the most critical step that assures the quality of the purified RNA and success of the subsequent RT-qPCR. The cells and tissues should be transferred into the lysis buffer as quickly as possible. When isolating RNA from

tissue blocks, sectioning times should be minimized and the tissue kept cold, preferably on dry ice. If the purified RNA is substantially degraded, it indicates that either the tissue was not immediately snap-frozen after dissection, freeze-thawed during storage, or thawed prior to contact with the lysis buffer.

3. RNA purification can also be done using Trizol reagent from Invitrogen, or similar. In this case, Trizol is added directly to the cells on the plate after removal of growth medium and PBS rinse. Tissue blocks can be either cryosectioned or ground in liquid nitrogen with a mortar and pestle, and Trizol is added directly to the frozen sections or powder. The mixture is transferred into an Eppendorf tube, vortexed, and RNA is isolated according to Invitrogen's instructions.
4. The success of the RNA purification step is determined by agarose gel electrophoresis. If RNA is intact, two major bands corresponding to 18S rRNA and 28S rRNA are detected. The presence of products below 50 bp is indicative of RNA degradation. If the RNA is not substantially degraded, increasing its amounts in subsequent reactions might be suitable for some applications. However, it is not applicable when the presence of large intact cDNA is needed, which is required, for example, for analysis of splice variants abundance.
5. The presence of a smear higher than 10 kb indicates that the purified RNA is contaminated with genomic DNA. In some cases, treatment with RNase-free DNase could be sufficient to remove traces of gDNA contamination. A control PCR reaction should be carried out using purified RNA as a template, to assure that all gDNA is removed.
6. The reverse-transcription reaction can be carried out using either random hexamers or oligo dTs as described here in this protocol. Priming with random hexamers is preferable when equal representation of 5′ and 3′ ends of the mRNA is crucial. This is especially important when analyzing splice variants of long mRNAs (21).
7. Set up of reactions for real-time PCR should be done very carefully and precisely. Amplification curves in parallel reactions should be nearly identical. Necessary precautions should be taken to avoid contamination with PCR products, such as separate location and pipettes assigned exclusively for PCR set up. Some labs prefer to have a special laminar cabinet, with a set of pipettes, devoted solely to PCR assembly. In such arrangement, the area can be easily decontaminated from any trace of PCR products by UV-irradiation.
8. When all PCR conditions are worked out, a large volume of Master Mix, which omits the DNA template, and Taq

polymerase can be premade and stored at −20°C in appropriate aliquots, i.e., sufficient for triplicate reactions.

9. Occasionally selected pair of primers and/or conditions are not optimal for successful amplification. To test this, PCR products need to be analyzed on an agarose gel, taking all precautions to avoid their spreading to the space, where PCR reactions are assembled. Only bands specific to the products of interest should be detected. In the case of nonspecific amplification, several steps can be undertaken to optimize the primers' performance. As a starting point, annealing temperature can be increased and the amount of primers or Mg^{2+} concentration in the reaction decreased. Specificity of PCR can also be greatly assisted by using "step down" method, where initial PCR annealing temperatures are several (4–6°C) degrees higher than the melting temperature of primers, and are then reduced stepwise to the melting point (22). Another cause of low PCR yield can be the innate complexity of the amplified region, such as high GC content and/or the presence of stable secondary structures. Amplification of such regions can be greatly assisted by addition of 1–2 M betaine to the PCR reaction.
10. Efficiency of qPCR reactions should be also tested by comparing amplification curves obtained using different dilutions of the same template. The threshold cycle C(T) for each amplification is calculated and plotted against template concentration log[C]. This should result in a linear plot with slope determining efficiency of amplification $E = 10^{[-1/\text{slope}]}$. Ideally, E should be equal 2, since during each PCR cycle the amount of DNA is doubled. If E is less than 2, amplification is not complete at each cycle. Value of E greater than 2 indicates excessive amplification, possibly due to the formation of additional nonspecific products.
11. The real-time PCR is often used to compare changes in gene expression induced upon drug treatment. The comparative C(T) method, also termed $2^{-\Delta\Delta C(T)}$ method, is used to analyze relative changes in gene expression in such experiments. The $\Delta\Delta C(T)$ is calculated as the difference in $\Delta C(T)$ for the treated and untreated samples. For the $\Delta\Delta C(T)$ calculations to be valid, amplification efficiencies for the reference and the tested gene must be approximately equal.

Acknowledgments

Authors would like to thank Alexander Skoblov and Michelle Nyhan for their help and advice in preparation of the manuscript.

References

1. Murphy, J. and Bustin, S.A. (2009) Reliability of real-time reverse-transcription PCR in clinical diagnostics: gold standard or substandard? *Expert Review of Molecular Diagnostics*, **9**, 187–197.
2. Schmittgen, T.D., Lee, E.J., Jiang, J., Sarkar, A., Yang, L., Elton, T.S. and Chen, C. (2008) Real-time PCR quantification of precursor and mature microRNA. *Methods (San Diego, Calif)*, **44**, 31–38.
3. Koga, T., Tokunaga, E., Sumiyoshi, Y., Oki, E., Oda, S., Takahashi, I., Kakeji, Y., Baba, H. and Maehara, Y. (2008) Detection of circulating gastric cancer cells in peripheral blood using real time quantitative RT-PCR. *Hepato-gastroenterology*, **55**, 1131–1135.
4. Rizzi, F., Belloni, L., Crafa, P., Lazzaretti, M., Remondini, D., Ferretti, S., Cortellini, P., Corti, A. and Bettuzzi, S. (2008) A novel gene signature for molecular diagnosis of human prostate cancer by RT-qPCR. *PloS One*, **3**, e3617.
5. Pignot, G., Bieche, I., Vacher, S., Guet, C., Vieillefond, A., Debre, B., Lidereau, R. and Amsellem-Ouazana, D. (2009) Large-scale real-time reverse transcription-PCR approach of angiogenic pathways in human transitional cell carcinoma of the bladder: identification of VEGFA as a major independent prognostic marker. *European Urology*, **56**, 678–688.
6. Andreeff, M., Ruvolo, V., Gadgil, S., Zeng, C., Coombes, K., Chen, W., Kornblau, S., Baron, A.E. and Drabkin, H.A. (2008) HOX expression patterns identify a common signature for favorable AML. *Leukemia*, **22**, 2041–2047.
7. Wakeley, P.R., Johnson, N., McElhinney, L.M., Marston, D., Sawyer, J. and Fooks, A.R. (2005) Development of a real-time, TaqMan reverse transcription-PCR assay for detection and differentiation of lyssavirus genotypes 1, 5, and 6. *Journal of Clinical Microbiology*, **43**, 2786–2792.
8. Lu, X., Holloway, B., Dare, R.K., Kuypers, J., Yagi, S., Williams, J.V., Hall, C.B. and Erdman, D.D. (2008) Real-time reverse transcription-PCR assay for comprehensive detection of human rhinoviruses. *Journal of Clinical Microbiology*, **46**, 533–539.
9. Wang, L., Giannoudis, A., Lane, S., Williamson, P., Pirmohamed, M. and Clark, R.E. (2008) Expression of the uptake drug transporter hOCT1 is an important clinical determinant of the response to imatinib in chronic myeloid leukemia. *Clinical Pharmacology and Therapeutics*, **83**, 258–264.
10. Mischitelli, M., Fioriti, D., Anzivino, E., Bellizzi, A., Ferretti, G., Gussman, N., Mitterhofer, A.P., Tinti, F., Barile, M., Dal Maso, M. *et al.* (2007) BKV QPCR detection and infection monitoring in renal transplant recipients. *New Microbiol*, **30**, 271–274.
11. Mackay, J. and Landt, O. (2007) Real-time PCR fluorescent chemistries. *Methods in Molecular Biology (Clifton, N.J)* **353**, 237–261.
12. Gibson, U.E., Heid, C.A. and Williams, P.M. (1996) A novel method for real time quantitative RT-PCR. *Genome Research*, **6**, 995–1001.
13. Tyagi, S. and Kramer, F.R. (1996) Molecular beacons: probes that fluoresce upon hybridization. *Nature Biotechnology*, **14**, 303–308.
14. Whitcombe, D., Theaker, J., Guy, S.P., Brown, T. and Little, S. (1999) Detection of PCR products using self-probing amplicons and fluorescence. *Nature Biotechnology*, **17**, 804–807.
15. Thelwell, N., Millington, S., Solinas, A., Booth, J. and Brown, T. (2000) Mode of action and application of Scorpion primers to mutation detection. *Nucleic Acids Research*, **28**, 3752–3761.
16. Lee, S., Gang, J., Jeon, S.B., Choo, S.H., Lee, B., Kim, Y.G., Lee, Y.S., Jung, J., Song, S.Y. and Koh, S.S. (2007) Molecular cloning and functional analysis of a novel oncogene, cancer-upregulated gene 2 (CUG2). *Biochemical and Biophysical Research Communications*, **360**, 633–639.
17. Kim, H., Lee, M., Lee, S., Park, B., Koh, W., Lee, D.J., Lim, D.S. and Lee, S. (2009) Cancer-upregulated gene 2 (CUG2), a new component of centromere complex, is required for kinetochore function. *Molecules and Cells*, **27**, 697–701.
18. de Kok, J.B., Roelofs, R.W., Giesendorf, B.A., Pennings, J.L., Waas, E.T., Feuth, T., Swinkels, D.W. and Span, P.N. (2005) Normalization of gene expression measurements in tumor tissues: comparison of 13 endogenous control genes. *Laboratory Investigation; a Journal of Technical Methods and Pathology*, **85**, 154–159.
19. Gresner, P., Gromadzinska, J. and Wasowicz, W. (2009) Reference genes for gene expression studies on non-small cell lung cancer. *Acta Biochimica Polonica*, **56**, 307–316.
20. Pernot, F., Dorandeu, F., Beaup, C. and Peinnequin, A. (2010) Selection of reference genes for real-time quantitative reverse transcription-polymerase chain reaction in hippocampal structure in a murine model of

temporal lobe epilepsy with focal seizures. *Journal of Neuroscience Research*, **8**, 1000–1008.

21. Taveau, M., Stockholm, D., Spencer, M. and Richard, I. (2002) Quantification of splice variants using molecular beacon or scorpion primers. *Analytical Biochemistry*, **305**, 227–235.
22. Roux, K.H. (2002) Single-step PCR optimization using touchdown and stepdown PCR programming. *Methods in Molecular Biology (Clifton, N.J.)*, **192**, 31–36.

Chapter 2

Reverse-Transcriptase Polymerase Chain Reaction to Detect Extracellular mRNAs

Sweta Rani and Lorraine O'Driscoll

Abstract

The presence of extracellular nucleic acids has been reported in serum/plasma from cancer and diabetes patients that may help in disease diagnosis. Taking insulin-producing cells as examples here, RT-PCR was used to investigate a correlation between the presence and amounts of extracellular mRNA(s) and cell mass and/or function. RT-PCR was performed on a range of mRNAs, including *Pdx1*, *Npy*, *Egr1*, *Pld1*, *Chgb*, *InsI*, *InsII*, and *Actb* in biological triplicate analyses.

Reproducible amplification of these mRNAs from MIN6, MIN6 B1, and Vero-PPI cells and their CM suggests that beta cells transcribe and release these mRNAs into their environment. mRNAs secreted from insulin-producing cells into their extracellular environment may have potential as extracellular biomarkers for assessing beta cell mass and function.

Key words: Extracellular nucleic acid, RT-PCR, Insulin-producing cells, Conditioned media, MIN6, Beta cell function, Beta cell mass

1. Introduction

In 1948, Mandel and Metais first reported detection of nucleic acids in plasma (1). Extracellular nucleic acid presents a good potential to be used as a tool for early diagnosis of disease. Extracellular RNA is detected in the serum and plasma of various forms of cancer and diabetes patients (2–4). DNA and mRNA from foetal origin have been discovered in the plasma of pregnant women (5–7).

There are many laboratory techniques developed to evaluate mRNA levels. The most commonly used technique to amplify nucleic acid is the reverse-transcriptase polymerase chain reaction (RT-PCR). This is a very sensitive method, and the genes expressed

Lorraine O'Driscoll (ed.), *Gene Expression Profiling: Methods and Protocols*, Methods in Molecular Biology, vol. 784,
DOI 10.1007/978-1-61779-289-2_2,

in very low levels are also detectable after exponential amplification. For this study, medium conditioned by a range of insulin-producing cell types, including glucose-responsive and nonresponsive murine beta cells, MIN6 (L) and (H), respectively, a glucose-responsive clonal population of MIN6 (MIN6 B1), and monkey kidney fibroblast cells engineered to produce human insulin (Vero-PPI) were used. Vero-PPI cells were previously engineered to produce human (pro)insulin (8). MIN6 B1 was seeded at a range of densities (1×10^6, 5×10^6, and 1×10^7) to investigate if the abundance of specific mRNAs detectable in conditioned media (CM) reflected the cell numbers conditioning the medium. A number of transcripts, including pancreatic and duodenal homeobox gene-1 (*Pdx1*), early growth response gene 1 (*Egr1*), chromagranin B (*Chgb*), insulin I (*InsI*), insulin II (*InsII*), neuropeptide Y (*Npy*), phospholipase D1 (*pld1*), and paired box transcription factor 4 (*Pax4*) were investigated.

1.1. Polymerase Chain Reaction

PCR was invented by Kary Mullis, and was awarded Nobel Prize for this (9). The word "polymerase" is derived from the enzyme DNA polymerase that plays an important role in copying DNA during replication or mitosis. PCR is the molecular biology technique used to exponentially amplify a DNA template using a thermal cycler. Selective and repeated amplification of the cDNA is performed using primer (consisting of complimentary sequences to the target) and DNA polymerase enzyme. *Taq* polymerase is the DNA polymerase enzyme isolated by Thomas D. Brock in 1976 from the thermophilic bacterium *Thermus aquaticus*, hence the name "*Taq*" polymerase (10).

1.2. Reverse Transcription

Howard Temin and David Baltimore shared the 1975 Nobel Prize in Physiology or Medicine with Renato Dulbecco for their discovery of an enzyme to transcribe DNA from RNA. PCR can only be used on DNA strands, but with the discovery of reverse-transcriptase, analysis of RNA molecules using PCR is possible. Synthesis of RNA from DNA is termed as "transcription." RT-PCR is reverse of "transcription" hence the term "reverse-transcription."

Reverse-transcriptase is a polymerase enzyme that transcribes single-stranded RNA into single-stranded DNA called complementary DNA or cDNA using oligo dT primer, gene-specific primers or random oligomers. Reverse transcription reaction could be carried out using total cellular RNA or poly(A) RNA, reverse-transcriptase enzyme, and primers, and can be converted into cDNA for further amplification using PCR. mRNA is isolated from tissues, cells, serum, CM, or any other desired samples and RT reaction is carried out using these mRNA. This chapter deals with performing RT-PCR using cell and CM RNA. CM is the medium that had been conditioned by cells for 48 h (see Note 1).

2. Materials

2.1. Cell Culture

1. MIN6 B1 cell line – DMEM media (stored at 4°C) with 150 mL/L heat-inactivated foetal calf serum (FCS) (stored at –20°C) (see Note 2), 2 mM L-Glut (stored at –20°C), and 75 µM ß-mercaptoethanol (store at room temperature). ß-Mercaptoethanol should be prepared in flow cabinet and filter sterilised before use. It may be stored at 4°C for a month. The medium could be prepared as required and must be used within 1 month.
2. MIN6 cell line – DMEM containing 25 mM glucose, supplemented with 200 mL/L heat-inactivated FCS (see Note 2).
3. Vero-PPI cell line – modified Eagle's medium (MEM) (store at 4°C) with 5.6 M glucose and 10 g/L non-essential amino acids (store at –20°C).
4. Solution of trypsin (0.25%) (store at –20°C) and ethylenediaminetetraacetic acid (EDTA) (1 mMol) (prepare, autoclave, and store at room temperature).
5. Phosphate-buffered saline (PBS): sodium chloride 8 g/L, potassium chloride 0.2 g/L, di-sodium hydrogen phosphate 1.15 g/L, potassium dihydrogen phosphate 0.2 g/L, pH 7.3 at 25°C, prepare and store at room temperature.

2.2. Isolation of RNA from Cells and CM

1. 0.45-µm filters (Millipore).
2. TRI Reagent (Sigma-Aldrich) should be used in fume hood; store at 4°C.
3. Chloroform (Sigma-Aldrich) should be used in fume hood; store at room temperature.
4. Glycogen (Sigma-Aldrich): final concentration 120 µg/mL (store at –20°C for up to 1 year).
5. Isopropanol (Sigma-Aldrich): store at room temperature.
6. Ethanol: make 75% using UHP and store at room temperature (For UHP see Note 3).
7. RNase-free water (Ambion); store at room temperature.

2.3. DNase Treatment of RNA

1. RNase-free DNase (Promega) (store at –20°C).
2. Reaction buffer: supplied as 10× – 400 mM Tris–HCl (pH 8.0), 100 mM $MgSO_4$, 10 mM $CaCl_2$ (Promega) (store at –20°C).
3. DNase Stop Solution: 20 mM EGTA (pH 8.0) (Promega) (store at –20°C).

2.4. RNase Treatment of RNA

1. RNase ONE Ribonuclease (Promega) (store at –20°C).
2. Reaction buffer: supplied as 10× – 100 mM Tris–HCl (pH 7.5 at 25°C), 50 mM EDTA, 2 M sodium acetate (Promega) (store at –20°C).

2.5. Reverse Transcription

1. Oligo dT primers (MWG); store at −20°C.
2. Human RNase inhibitor (Sigma-Aldrich); store at −20°C.
3. Moloney murine leukaemia virus (MMLV) reverse-transcriptase (RT) enzyme (Sigma-Aldrich); store at −20°C.
4. Reverse-transcriptase buffer: supplied as 10× – 500 mM Tris–HCl (pH 8.3), 500 mM KCl, 30 mM $MgCl_2$, 50 mM DTT; store at −20°C.

2.6. PCR Analysis

1. $MgCl_2$: supplied with *Taq* polymerase enzyme (Sigma-Aldrich; store at −20°C).
2. Oligonucleotide primers, see Note 4 (MWG; store at −20°C).
3. *Taq* DNA polymerase enzyme (Sigma-Aldrich; store at −20°C).
4. Deoxynucleoside triphosphate (dNTP) (Sigma-Aldrich; store at −20°C).

2.7. Electrophoresis

1. Agarose (Sigma-Aldrich); store at room temperature.
2. Tris-borate-EDTA (TBE) buffer: 10.8 g/L Tris base; 5.5 g/L boric acid, 4 mL/L 0.5 M EDTA (pH 8.0). Prepare in UHP and store at room temperature.
3. Ethidium bromide, a carcinogen, should be used with caution (Sigma-Aldrich; (10 mg/mL)), prepare and store at room temperature.
4. ΦX174 DNA HaeIII digest (Sigma-Aldrich; stored at −20°C).
5. 6× loading buffer: 50% glycerol, 1 mg/mL bromophenol blue, 1 mM EDTA; prepare in UHP and store at room temperature.

2.8. Equipments Required

1. Thermal cycler.
2. NanoDrop.
3. LabWorks Analysis Software (version 3.0; UVP).
4. Electrophoresis unit (Bio-Rad).
5. EpiChemi II Darkroom, UVP Laboratory Products.

3. Methods

RNA from cells and their CM are isolated using TRI Reagent and are quantified using NanoDrop. RNA is easily degraded by ubiquitously present RNase (ribonuclease) enzymes, so precaution should be taken prior to RNA work. All Eppendorfs, PCR tubes, Gilson pipette tips, etc., should be RNase-free, and disposable nitrile gloves should be worn.

Proper controls should be included to ensure that the amplified product is of RNA origin, and not from contaminating genomic DNA. Some of the relevant controls to include involve analysing samples after DNase treatment; after RNase treatment; and without reverse-transcriptase enzyme – an enzyme which is necessary for the formation of cDNA on an mRNA template. Water, on its own, should also be included as a control to determine the presence of any contaminant (PCR-VE and RT-VE). Treatment with DNase enzyme would destroy any contaminating DNA leaving only RNA, with RNase enzyme would destroy all RNA leaving only genomic DNA if any (see Note 5).

3.1. Preparation of Conditioned Media and Cells

1. Cells are grown to ~80% confluency, re-feeding every 3 days. To condition the medium for analysis, fresh medium (8 mL) is added for a further 48 h (by which time the cells have reached ~90% confluency).
2. CM is passed through a 0.45-μm filter to ensure that no cells or large cell particles are present.
3. 250 μL aliquots of filtered CM is added to 750 μL of TRI Reagent and incubated at room temperature for 5–10 min to completely dissociate nucleoprotein complexes. Aliquots are stored at −80°C until RNA extraction and analysis.
4. The corresponding cells from the flasks are washed with PBS, trypsinised, and centrifuged at 7,500 × *g* for 5 min. The pellets are then washed twice with cold PBS, resuspended in 1 mL TRI Reagent, incubated as described above, and stored at −80°C.

3.2. Isolation of RNA from Cells and CM

1. The frozen TRI Reagent samples are allowed to thaw at room temperature. Upon thawing, allow these to sit for at least 5 min for complete dissociation of nucleoprotein complexes.
2. To this, add 0.2 mL of chloroform per mL of TRI Reagent. Shake samples vigorously for 15 s and allow to stand for 15 min at room temperature (see Note 7).
3. The resulting mixture is then centrifuged at 15,700 × *g* in a microfuge for 15 min at 4°C.
4. The colourless upper aqueous phase (containing RNA) is removed into a fresh RNase-free 1.5-mL Eppendorf tube.
5. To this tube, add 1.25 μL of glycogen (added only to the CM samples) and 0.5 mL of ice-cold isopropanol. Mix the samples, incubate at room temperature for 5–10 min, and store at −20°C overnight to ensure maximum RNA precipitation (see Note 6).
6. To pellet the precipitated RNA, centrifuged the Eppendorf tubes at 13,400 × *g* for 30 min at 4°C.
7. Taking care not to disturb RNA pellet, remove the supernatant. Wash the pellet by the addition of 750 μL of 75% ethanol and vortex. Centrifuge at 5,400 × *g* for 5 min at 4°C.
8. Repeat step 7.
9. The RNA pellet is then allowed to air-dry for 5–10 min, and is then resuspended in 15 μL of RNase-free water. To facilitate dissolution of the RNA pellet repeated pipetting may be included.

3.3. DNase Treatment of RNA

To digest any contaminating genomic DNA from RNA isolates, cell and CM samples are treated with RNase-free DNase as follows:

1. 1 μg and 4 μL of RNA isolate is taken from cell and CM samples, respectively.
2. To this, add 1 U RNase-free DNase.
3. Add 1 μL of reaction buffer to this mixture.
4. Then add RNase-free water, bringing the volume up to 10 μL.
5. This mixture is then incubated at 37°C for 30 min.
6. The DNase is then inactivated by adding 1 μL DNase Stop Solution.
7. The reaction mixture is incubated at 65°C for 10 min.

3.4. RNase Treatment of RNA

To digest all RNAs present in aliquots from all cell and CM samples, treat with RNase ONE Ribonuclease as follows:

1. RNA isolates are treated with 1 U RNase ONE Ribonuclease per 0.1 μg RNA.
2. To this, add 10 μL of reaction buffer.
3. This mixture is further incubated at 37°C for 30 min to destroy all RNA present.

3.5. Reverse Transcription

1. First-strand cDNA is synthesised using 1 μL of 500 ng/L oligo dT primers per 1 μg RNA from cultured cells and 4 μL of the RNA suspension from CM (see Note 8).
2. RNase-free water is added to make up the final volume to 5 μL and should be incubated at 72°C for 10 min.
3. This mixture is then cooled on ice, and the following reaction mixture is added to make up 15 μL of final volume:

   1 μL of human RNase inhibitor, 10 mM of each (dNTP, 1 μL of MMLV (200 U/μL) reverse-transcriptase, 2 μL of 10× reverse-transcriptase buffer, and RNase-free water.
4. This mixture is incubated at 37°C for 1 h.
5. RT reaction is subsequently set up for all the controls using aliquots of isolates treated with DNase enzyme; with RNase enzyme; or untreated (see Note 5).
6. In parallel, an RT reaction is set up without RT enzyme (MMLV-RT) as control (see Note 5).

3.6. PCR Analysis

1. cDNAs (2.5 μL) from CM and cells are amplified in a 25 μL PCR reaction solution containing 1.5 mM $MgCl_2$, 0.2 mM of each dNTP, 20 μM oligonucleotide primers, and 2.5 U *Taq* polymerase enzyme.
2. To detect if mRNAs detected extracellular to cells are likely to be full-length transcripts or if they are fragmented products, PCR primers are designed in such a way as to amplify regions

close to the 3′ end; close to the 5′ region; or stretching along most of the length of the transcript (see Note 4).

3. The mixture is then amplified using following conditions.

95°C for 3 min	(Denaturation)
Followed by 30–45 cycles of (see Note 8): 95°C for 30 s 52–60°C for 30 s 72°C for 45 s	 (Denaturation) (Annealing) (Extension)
And 72°C for 10 min	 (Extension)

3.7. Gel Electrophoresis of PCR Products

1. The PCR-amplified products are subsequently separated by electrophoresis on a 2% agarose gel prepared in a 1× TBE buffer by melting in a laboratory microwave. Thick gloves and suitable face-protecting safety gear should be used when handling boiling agarose.
2. Upon cooling, the gel is supplemented with 5 μL ethidium bromide (10 mg/mL) to allow visualisation of the DNA upon intercalation.
3. The gel is then poured in to the electrophoresis unit and allowed to set. Sample wells are formed by placing a comb into the top of the gel prior to its setting.
4. To run the samples, 2 μL of 6× loading buffer is added to 10 μL of PCR product and the mixture is loaded to the gel with an appropriate size marker (e.g. ΦX174 DNA HaeIII digest).
5. Gels are electrophoresed at 120–150 V for 1–2 h (depending on size of the target gene, i.e. to get adequate separation).
6. Once the internal control and target bands have migrated to the required extent, the gel is taken to the gel analyzer (EpiChemi II Darkroom, UVP Laboratory Products), photographed, and densitometrically analysed using Labworks software (UVP).

3.8. Densitometry Analysis

1. Densitometric analysis of the PCR products may be performed using the MS Windows 3.1 compatible Molecular Analyst software/PC image analysis software available for use on the 670 Imaging Densitometer (Bio-Rad. CA) Version 1.3; or other suitable packages.
2. Developed negatives of gels are scanned using transmission light, and the image is transferred to the computer.
3. The amount of light blocked by the DNA band is in direct proportion to the intensity of the DNA present. A standard area is selected and scanned. A value is taken for the optical density (OD) of each individual pixel on the screen.

4. The average value of this OD (within a set area, usually cm^2) is normalised for the background of an identical selected area.
5. The normalised reading is taken as the densitometric value used in analysis. These OD readings are unit less, i.e. are assigned arbitrary units.
6. The results are imported into Microsoft Excel, and bar charts may be generated from this data.

3.9. Results

1. While 30 cycles of PCR is found to be adequate for analysis of transcripts using RNA isolated from cell lines, often products are undetectable or very low in intensity when analysing RNA from CM after 30 cycles of PCR. Forty-five cycles of PCR, however, is generally adequate to produce a detectable band from CM RNA (11) (see Note 9).
2. Amplified products are obtained from cell and CM isolates that were untreated with a digestive enzyme or that were treated with DNase enzyme. However, no bands were detected where samples had been treated with RNase enzyme (see Note 5; Fig. 1).
3. The absence of RT enzyme (MMLV-RT) results in no amplified PCR products, regardless of whether the cell and CM isolates were untreated, RNase- or DNase treated (Fig. 1).
4. In the case of chromogranin b, the intensity of the band produced following 45 cycles of PCR is directly associated with the numbers of cells conditioning the medium. Increased

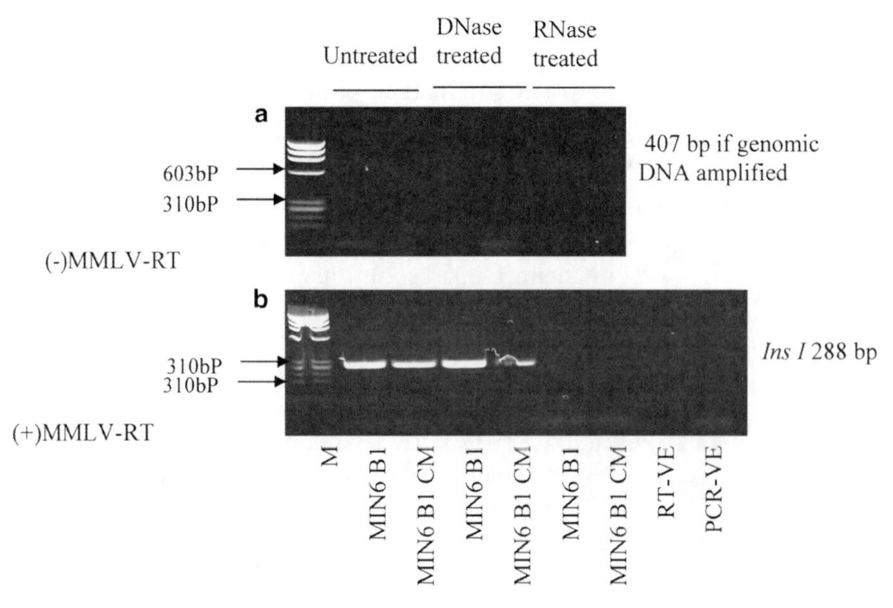

Fig. 1. Cell and CM RNA isolates treated with RNase or DNase prior to cDNA formation, using MMLV-RT enzyme, and subsequent amplification using *InsI* primers. (**a**) (+) MMLV-RT = reverse-transcriptase (RT) reaction performed with necessary RT enzyme, MMLV-RT. (**b**) (–) MMLV = RT cycle performed in the absence of MMLV-RT enzyme, as control. *RT-VE* = reaction with H_2O instead of RNA as control; *PCR-VE* = PCR reaction with H_2O instead of cDNA as a control. *M* = molecular weight marker: ΦX174 DNA HaeIII digest.

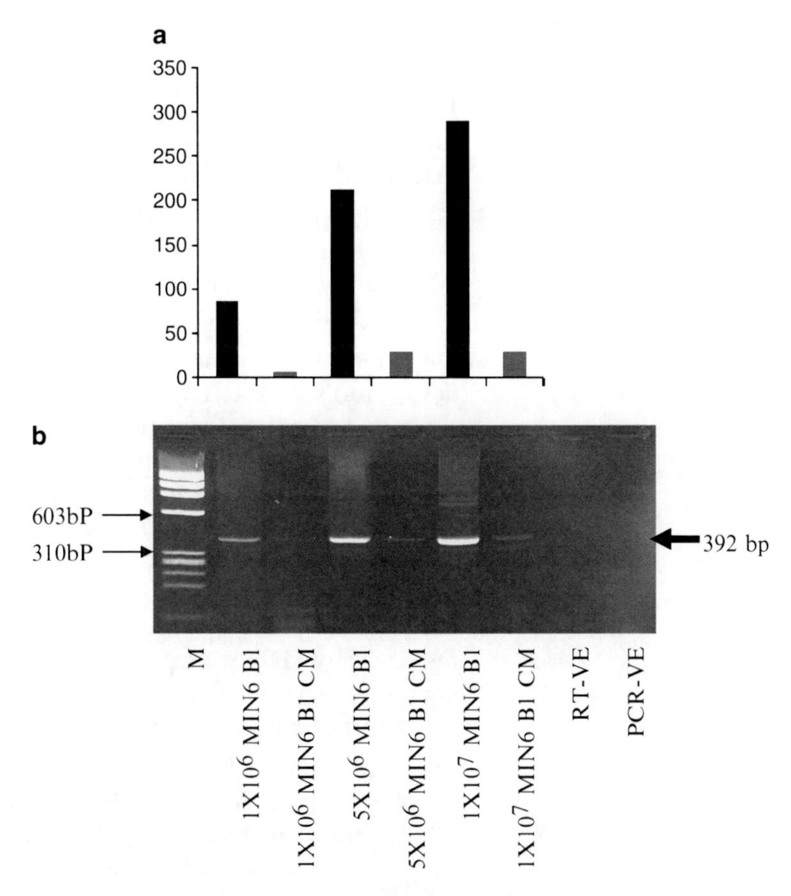

Fig. 2. (**a**) Densitometric analysis (**b**) *Glucokinase* RT-PCR on RNA isolated from MIN6 B1 cells, grown at a range of densities, and their corresponding CM. *RT-VE*=reaction with H_2O instead of RNA as control; *PCR-VE*=PCR reaction with H_2O instead of cDNA as control. Densitometry analysis of amplified gene products is shown above the gel image. *M*=molecular weight marker: ΦX174 DNA HaeIII digest. Results are representative of $n=3$ biological repeats.

number of cells resulted in increased levels of these transcripts detectable in a fixed volume of CM (Fig. 2).

4. Notes

1. To identify a suitable time-point at which mRNA could be routinely amplified, CM samples are collected at four time-points (24, 48, 72, and 96 h) after seeding cells. Beta-actin, a house keeping gene that is highly expressed gene transcript may be isolated and amplified at all time-points evaluated. However, for low levels of gene transcripts expressed by the cell populations, amplified product is undetected after 24 h, but is present when analysed after 48 h, or more, of culture.

To ensure that all CM analysed is from healthy, proliferating cells, it may be recommendable to analyse 48 h CM.

2. FCS should be thawed at 4°C overnight, or at room temperature, before heat-inactivating. Thawed serum may be heat-inactivated at 56°C for 1 h in a water bath.

3. Ultra high pure water (UHP) is the pretreatment of water involving activated carbon, pre-filtration and anti-scaling. Water is then purified by a reverse osmosis system (Millipore Milli-RO 10 Plus) and has a resistivity of 18.2 MΩ-cm.

4. Oligonucleotide primers may be self-designed using primer design tools or may be purchased pre-designed. It has been suggested that RNA detected in human serum/plasma probably are present as short fragments (12). In this particular study, as example, we amplified products ranging from 127 to 383 bp in length, at various regions (5′, 3′, and – in some cases – amplifying most of the sequence) along the length of the mRNAs (full-length transcripts ranging from 0.294 to 3.111 kb in size). We were able to successfully amplify all regions along the length of cDNAs prepared using oligo(dT) primers targeting the poly(A) tail of mRNAs, suggesting that the CM mRNAs are not fragmented, but are full-length products.

5. In a study using saliva specimens, it was suggested (13) that microarray and qRT-PCR analysis might be detecting genomic DNA, rather than mRNA (14). Therefore, proper controls must be included to determine if the extracellular nucleic acids detected are either wholly or partly DNA in origin, and not from mRNA. As controls, RT-PCR analysis following DNase treatment; RNase treatment; and in the absence of RT enzyme should be performed. Amplified products detected following DNase treatment of samples, complete lack of product following RNase treatment, and in the absence of RT enzyme, supports that the nucleic acids detected are of RNA, not DNA, origin. The amplified products detected should be of the size expected for cDNA, and not of genomic DNA.

6. Glycogen acts as a carrier or co-precipitant in RNA purification. Glycogen is added and left overnight at –20°C, to ensure maximum RNA precipitation. However, for isolating RNA from cells only, a glycogen carrier is unnecessary due to the substantially greater amounts of RNA present. Chloroform helps in phase separation of the mixture into three phases: a red organic phase containing protein, an interphase containing DNA, and an upper colourless aqueous phase containing RNA.

7. A constant amount of total RNA (e.g. 1 μg, as determined by NanoDrop) is used for all cell analysis. In the case of CM, the levels of mRNA are lower and it is likely that all types of total RNA detected in the cell are not present in CM. The numbers

of cells seeded should be accurately counted, and the volumes of medium used kept constant for all replicates. Therefore, for analysis of mRNA in CM a constant volume (e.g. 4 μL) of RNA suspension may be analysed.

8. 30 cycles of PCR is sufficient when amplifying gene transcripts from cell RNA. In some cases, very low intensity band or no bands are observed when CM RNA was amplified at 30 cycles of PCR. So it may be necessary to increase this to 45 cycles when using RNA extracted from CM.

References

1. Mandel, P., Metais, P. (1948) Nucleic acids of human blood plasma. *C R Seances Soc Biol Paris.* **142**, 241–243.
2. Wieczorek, A.J, Sitaramam, V., Machleidt, W., Rhyner, K., Perruchoud, A.P., Block, L.H. (1987) Diagnostic and prognostic value of RNA-proteolipid in sera of patients with malignant disorders following therapy: First clinical evaluation of a novel tumor marker. *Cancer Res.* **47**, 6407–6412.
3. Tsui, N.B., Dennis, Lo. Y.M. (2006) Placental RNA in maternal plasma: Toward noninvasive fetal gene expression profiling. *Ann N Y Acad Sci.* **1075**, 96–102.
4. Poon, L.L., Leung, T.N., Lau, T.K., Lo, Y.M. (2000) Presence of fetal RNA in maternal plasma. *Clin Chem.* **46**, 1832–1834.
5. Bianchi, D.W., Williams, J.M., Sullivan, L.M., Hanson, F.W., Klinger, K.W., Shuber, A.P. (1997) PCR quantitation of fetal cells in maternal blood in normal and aneuploid pregnancies. *Am J Hum Genet.* **61**, 822–829.
6. Hasselmann, D.O., Rappl, G., Rossler, M., Ugurel, S., Tilgen, W., Reinhold, U. (2001) Detection of tumor-associated circulating mRNA in serum, plasma and blood cells from patients with disseminated malignant melanoma. *Oncol Rep.* **8**, 115–118.
7. Kopreski, M.S., Benko, F.A., Kwak, L.W., Gocke, C.D. (1999) Detection of tumor messenger RNA in the serum of patients with malignant melanoma. *Clin Cancer Res.* **5**, 1961–1965.
8. O'Driscoll, L., Gammell, P., Clynes, M. (2002) Engineering vero cells to secrete human insulin. *In Vitro Cell Dev Biol Anim.* **38**, 146–153.
9. Mullis, K.B. (1990) The unusual origin of the polymerase chain reaction. *Sci Am.* **262**, 56–61.
10. Chien, A., Edgar, D.B., Trela, J.M. (1976) Deoxyribonucleic acid polymerase from the extreme thermophile *Thermus aquaticus. J Bacteriol.* **127**, 1550–1557.
11. Rani, S., Clynes, M., O'Driscoll, L. (2007) Detection of amplifiable mRNA extracellular to insulin-producing cells: potential for predicting beta cell mass and function. *Clin Chem.* **53**, 1936–44.
12. El-Hefnawy, T., Raja, S., Kelly, L., Bigbee, W.L., Kirkwood, J.M., Luketich, J.D., Godfrey, T.E. (2004) Characterization of amplifiable, circulating RNA in plasma and its potential as a tool for cancer diagnostics. *Clin Chem.* **50**, 564–573.
13. Kumar, S.V., Hurteau, G.J., Spivack, S.D. (2006) Validity of messenger RNA expression analyses of human saliva. *Clin Cancer Res.* **12**, 5033–5039.
14. Li, Y., Zhou, X., St John, M.A., Wong, D.T. (2004) RNA profiling of cell-free saliva using microarray technology. *J Dent Res.* **83**, 199–203.

Chapter 3

Microarray Analysis of mRNAs: Experimental Design and Data Analysis Fundamentals

Jai Prakash Mehta

Abstract

Microarray technology has made it possible to quantify gene expression of thousands of genes in a single experiment. With the technological advancement, it is now possible to quantify expression of all known genes using a single microarray chip. With this volume of data and the possibility of improper quantification of expression beyond our control, the challenge lies in appropriate experimental design and the data analysis.

This chapter describes the different types of experimental design for experiments involving microarray analysis (with their specific advantages and disadvantages). It considers the optimum number of replicates for a particular type of experiment. Additionally, this chapter describes the fundamentals of data analysis and the data analysis pipeline to be followed in most common types of microarray experiment.

Key words: Microarray, Gene expression, Experiment design, Normalisation, Clustering

1. Introduction

Expression microarray profiling is a high-throughput technology used in molecular biology to simultaneously access the gene expression profile of thousands of genes. A typical microarray chip consists of an arrayed series of thousands of microscopic spots of DNA oligonucleotides, each containing a small amount of a specific DNA sequence. This can be a short section of a gene or other DNA element that are used as probes to hybridise a cDNA or cRNA sample, under appropriate conditions. The hybridisation is detected and quantified by fluorescence-based detection of fluorophore-labelled targets, to determine relative abundance of nucleic acid sequences in the sample.

In standard microarrays, the probes are attached to a solid surface (made of glass or silicon) by a covalent bond to a chemical

Lorraine O'Driscoll (ed.), *Gene Expression Profiling: Methods and Protocols*, Methods in Molecular Biology, vol. 784, DOI 10.1007/978-1-61779-289-2_3,

matrix via epoxy-silane, amino-silane, lysine, and polyacrylamide (1). Affymetrix technology uses a photolithographic technology to synthesise 25-mer oligonucleotides on a silica wafer (http://www.affymetrix.com). Other microarray platforms, such as Illumina, use microscopic beads, instead of the large solid support (http://www.illumina.com/). While there are many options available, the Affymetrix chips are frequently used and so have been selected for more detailed consideration here.

1.1. Affymetrix Platform

The probes included on the Affymetrix platform are manufactured on the chip using photolithography (a process of using light to control the manufacture of multiple layers of material), which is adapted from the computer chip industry. Each GeneChip contains approximately 1,000,000 features. Each probe is spotted as a pair, one being a perfect match (PM) and the other with a mismatch (MM) at the centre. These probe pairs allow the quantification and subtraction of signals caused by non-specific cross-hybridisation. The differences in hybridisation signals between the partners, as well as their intensity ratios, serve as indicators of specific target abundance. Each gene or transcript is represented on the GeneChip by 11 probe pairs. The probe sets are given different suffixes to describe their uniqueness and/or their ability to bind different genes or splice variants.

- "_at" describes probes set that are unique to one gene.
- "_a_at" describes probe sets that recognise multiple transcripts from the same gene.
- "_s_at" describes probe sets with common probes among multiple transcripts from separate genes. The _s_at probe sets can represent shorter forms of alternatively polyadenylated transcripts, common regions in the 3′ ends of multiple alternative splice forms, or highly similar transcripts. Approximately 90% of the _s_at probe sets represent splice variants. Some transcripts are also represented by unique _at probe sets.
- "_x_at" designates probe sets, where it was not possible to select either a unique probe set or a probe set with identical probes among multiple transcripts. Rules for cross-hybridisation are dropped in order to design the _x_at probe sets. These probe sets share some probes with two or more sequences, and therefore these probe sets may cross-hybridise in an unpredictable manner.

Once the array is scanned, an image file is created called a ".dat" file. The software then computes cell intensity data (".cel" file) from the image file. It contains a single intensity value for each probe cell delineated by the grid (calculated by the Cell Analysis algorithm). The amount of light emitted at 570 nm from stained chip is proportional to the amount of labelled RNA bound to each probe. Each spot corresponds to individual probe (either perfect match or mismatch). The probes for each gene are distributed randomly across the chip to nullify any region-specific bias. Following this, data analysis algorithms combine the probes to the respective intensity of individual transcripts.

2. Experimental Design

An important consideration regarding experimental design is the microarray platform that has been selected to be used. Primarily, the experimental design depends on whether the experiment is a single channel experiment (i.e. only one dye has been used) or the experiment is a dual channel experiment (i.e. two dyes have been used; one for the sample and the other as a reference).

2.1. Double Channel Experiment

Double channel experiments include two dyes, i.e. one for each sample being compared. These are hybridised on the same chip, and the expression indices is the relative expression among the two samples that are hybridised onto a single chip. Some of the common types of design are as outlined below:

1. Common reference design: In such experiments, a pool of RNA is used as a reference sample, and all the samples are co-hybridised with the reference sample. The expression measure, therefore, is with respect to the reference design and such pairs of hybridisation can be compared across samples. The advantages of this design include both the relative ease of data analysis and the saving on microarrays chips – when compared to some other designs. Another advantage of such experiments is the ease with which the experiment can be extended. Reference design is by far the most common design performed in microarray experiment, due to the simplicity of the analysis of such types of data. However, it is argued that it is a very inefficient design because the reference sample, which is of least importance, hybridises half of the total hybridisations (Fig. 1) (2).
2. All combination design: In such experiments, all samples are hybridised – and so compared with – all samples (Fig. 2). The results obtained from such experiments are highly reliable.

Fig. 1. Example of common reference design.

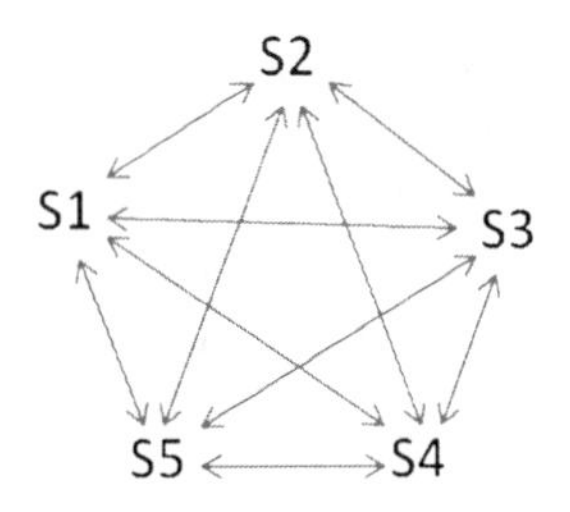

Fig. 2. Example of all combination design.

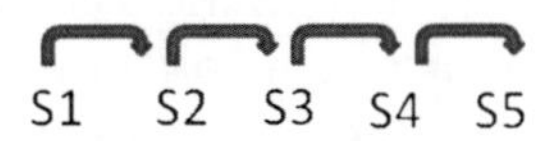

Fig. 3. Example of loop design.

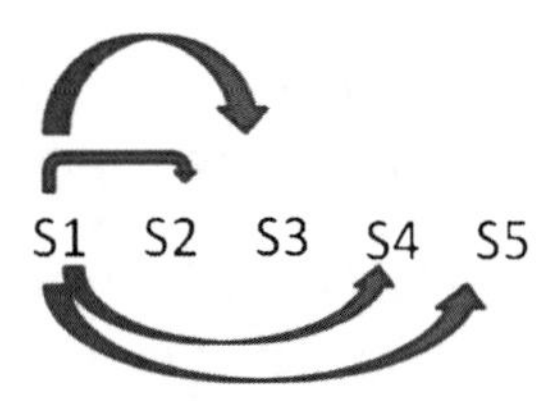

Fig. 4. Example of one sample as reference.

However, the cost of such experiments is very high if there is a large number of samples in an experiment.

3. Loop design: This design is more suited for time series experiment. In such a design, the first time point is hybridised to second; second with the third; and so on, thereby creating a loop (Fig. 3). Data analysis from such types of design may be complicated. The loop design is often considered superior to the reference design (3).
4. One sample as reference: In such experiments, one sample is used as a reference and all the samples are hybridised to that sample (Fig. 4). Such experimental designs are suited for time series experiments and often the sample at time 0 is taken as the reference sample.

2.2. Single Channel Experiment

For single channel experiments, only one dye is used and the expression measure is an absolute measure. This technology has been standardised by Affymetrix for their oligonucleotide arrays. Experimental design is very simple. There is no reference, and the controls and treatments are hybridised on separate chips. The only issue in such experimental design is the number of replicates.

2.3. Replicates

Replicates add to the cost of experiment and, in early days of microarray technology, it was often questionable as to whether or it is necessary to include replicates. Replicates, undoubtedly, make the results more reliable and less prone to misclassification (4). The next question, however, is how many replicates should be included? There is no thumb-rule to decide on this, and it should be decided based on the number of factors, some of that are listed below:

1. Sample quality: Some samples are "noisier" (have high variability among the replicates) than others, e.g. cell lines expression is considered be less noisy than that in tissues from patients as the RNA obtained can be from mixed cell types. Pooling of samples may reduce noise and would be specially suited if, the overall trend, rather than expression in individual specimens is required.
2. Expected effect: If the expected differences are large, sample size can be reduced. The replicate number can also be reduced if the aim is to get a holistic picture, e.g. pathways and ontology rather than expression of individual genes. However, if the expected differences are small, increasing the number of replicates is more desirable. Microarray expression values are a lot less reliable at low expression values, and so if the experimental objective includes assessing small differences in expression levels, a high number of replicates is desirable.
3. Experimental design: Experimental design dictates the number of replicates. All combination design needs more arrays; however, replicates of each of these may be unnecessary. Dye-swap experiment design, where the dyes allotment to each sample is reversed to counterbalance any dye-related bias, requires an additional replicates for each pair of hybridisation.
4. Microarray platform: The commercial microarray platforms, e.g. Affymetrix, Illumina, Agilent, etc., offer a high degree of reproducibility of data, and therefore technical replicates tend to be less critical than where in-house printed chips are used for the first time.
5. Biological compared to technical replicates: With advancing technology becoming quite robust, there may be more limited need for many technical replicate. The main focus should, therefore, be on the number of biological replicate to be included within the study.

Even with all the above knowledge, it can be difficult to predict the optimum numbers needed. Lee et al. (4), after performing a very tightly designed experiment, concluded that at least three biological replicates should be used for cDNA microarrays to get reliable result. Pavlidis et al. (5) concluded that at least five replicates should be used for any microarray experiment and that 10–15 replicates would yield results which are stable. In analysis involving clinical specimen, due to person-to-person variability, etc., a much higher number of replicates is needed compared to analysis of cell lines cultured under fixed conditions.

3. Data Analysis

3.1. Image Analysis and Quantification

After the hybridisation step, the chips are scanned and the intensity obtained is proportional to the expression measure. For a single channel experiment, such as Affymetrix, one scanning is required; for two-channel experiments, two scannings at different wavelength are required. For the commercially available chips, e.g. Affymetrix, the hardware comes with the required software to process the image to expression values.

For other types of chips, image analysis software, such as Spotfinder, may be used. The software imports the tiff image obtained after scanning the chip then performs semi-automated grid adjustment to identify individual spots followed by quantifying the intensity of the non-saturated pixels. Background correction is applied to get the desirable expression measure for each spot which represents the expression values for that transcript.

Such type of chips uses two dyes, one for the sample and the other for the reference. The samples are mixed before hybridisation on the chip. This gives a relative expression of sample with reference. However, owing to differential ability to detect the two dyes, the resultant expression is biased towards either of two dyes.

3.2. Normalisation

Normalisation is the first step in the data analysis process. Normalisation is the adjusting of microarray data to remove variations that arise from the technology rather than from biological differences between the RNA samples. These differences may be due to unequal quantity of starting RNA, differences in labelling or detection ability of the two dyes and any other systematic bias, such as print-tip, etc. Normalisation techniques differ significantly between two-channel and one-channel experiments.

Normalisation of two-channel experiment is simpler than that of one-channel experiment as the expression measures obtained is relative expression rather than absolute expression. Few of the normalisation techniques for cDNA arrays are discussed here.

1. Scaling: The central concept to this normalisation is that there are large number of genes in RNA and only a small portion of them are differentially expressed, and the average up-regulation is nearly same as average down-regulation. Considering these assumptions, the average intensity of the two-channels should be same, and if there is an overall high or low expression, then a scaling factor is calculated and multiplied by one or both to make the channel mean same.

   $\sum R = f \sum G$ (R indicates expression measure for one dye and G indicates expression measure of other dye). The factor (f) is calculated and multiplied by expression values of G. There are many variations to this method. Quiet often, the expression measures are Log 2 (Ratio). To achieve the same, the individual samples are mean centred, i.e. the mean for entire sample is calculated and is then subtracted from expression measure for each transcript.

2. Lowess normalisation: Lowess (locally weighted least square regression) normalisation is a normalisation technique which takes account the variation in the intensity detection at the local level rather than globally (6). To check for whether loess normalisation would be suited for the data a loess plot is constructed plotting $\log(R/G)$ versus $\log(\mathrm{sqrt}(R \times G))$. If the plots follow a curve, loess normalisation would be suited for the data. This normalisation works on an assumption that dye bias is dependent on spot intensity. Thus, this normalisation technique fits a locally weighted least squares regression to the data using a window frame defined by the user. A large window frame would give a smoother curve, whereas a smaller window frame would results in more local variation.

   Therefore, the new expression measure would be

$$\mathrm{Log}\left(R / G\right) \Rightarrow \log\left(R / G\right) - c(A)$$

   where $c(A)$ is the Lowess fit to the $\log(R/G)$ versus $\log(\mathrm{sqrt}(R \times G))$ plot.

3. Affymetrix data normalisation

   Some of the common normalisation algorithms for Affymetrix arrays are MAS5 (7), RMA (8), and dCHIP (9).

   MAS5 developed by Affymetrix uses a reference (baseline) chip, which is used to normalise all the experimental chips. The procedure is to construct a plot of each chip's probes against the corresponding probes on the baseline chip (after eliminating the highest 1% and lowest 1% probes). A regression line is fitted on the middle of the remaining 98% of the probes and based on this the target chip is normalised.

   dCHIP uses an array with median overall intensity as the baseline array against which other arrays are normalised at probe level intensity. Subsequently, a subset of PM probes, with small within-subset rank difference in the two arrays (also known as invariant set), serves as the basis for fitting a normalisation curve.

   RMA employs normalisation at probe level using the quantile method. This normalisation method makes the chips have identical intensity distribution.

3.3. Differentially Expressed Genes

Microarray experiments are primarily performed to identify differentially expressed genes. Various methods have been developed to identify differentially expressed genes in a microarray experiment.

Some of the basic approaches that are used to identify differentially expressed genes are described below.

1. Fold change
   Fold change is the ratio of the mean of the experimental group to that of the baseline. It is a metric to define the gene's mRNA-expression level between two distinct experimental conditions. Earlier, when the microarray experiments were costly and replicates were not possible, fold change was the only option to identify the differentially expressed genes. In the absence of replicates, the fold change cut-off is taken as 2. However, with the technology getting cheaper, more and more studies are going with high number of replicates. Therefore, fold change filtration can be applied in combination with other statistical criteria, such as *t*-test, etc. In such cases, fold change cut-off can be reduced to 1.5 or even 1.2.
2. Difference
   The difference of expression units can also be incorporated for finding differentially regulated genes. For log-ratio expression measure, difference of expression is considered as fold change, and therefore more commonly used for cDNA microarrays.
3. *t*-test
   The *t*-test analyses the mean and deviation of the samples and control distribution and calculates the probability that the observed difference in mean occurs when the null hypothesis is true. The null hypothesis states that the means of the two distributions is equal. Therefore, the hypothesis testing allows calculating the probability of finding the observed data when the hypothesis is true. For calculation of the probability, normal distribution is used. When the probability is low, the null hypothesis is rejected.

   p-value is the probability, if the test statistics were distributed as it would be under the null hypothesis, of observing a test statistic as extreme as, or more extreme than the one actually observed. For example, a *p*-value of 0.05 indicates that there would be a 5% chance of drawing the sample being tested, if the null hypothesis was true. A *p*-value of 0.05 is a typical threshold used in most of the situations to evaluate the null hypothesis. A more stringent lower *p*-value may be applied to obtain reliable result.

   The *t*-test is calculated as follows:

   $$t = (X_T - X_C) / \sqrt{(\mathrm{var}_T / n_T + \mathrm{var}_C / n_C)}$$

   X_T is the mean of Treatment, X_C is the mean of Control, var_T is the variance of Treatment, var_C is the variance of Control, n_T is the number of Treatment, and n_C is the number of Control.

4. False discovery rate
   False discovery rate (FDR) is an estimate of calling a gene as significantly differentially expressed when it is actually not differentially expressed. Since thousands of genes are compared, chances are that lot of genes can come as significant just by chance. Permutation of the samples in random groups and estimating the number of significant genes is one way of getting an estimate of FDR. In this process, the two groups are assigned with random samples and the numbers of differentially expressed genes are calculated. The process is repeated a number of times (higher the repeats, more accurate would be the estimate of FDR) and a median FDR is calculated. The FDR can vary depending on the experimental setup, the type of samples and the difference among the groups. The aim of the filtration criteria should be to obtain a low FDR. An FDR around 5–10% is normally acceptable. Certain algorithm identifies differentially expressed genes by adjusting the *p*-value cut-off and by controlling the FDR (10).

3.4. Clustering

Clustering is the grouping of objects based on similarity. In other words, it is the partitioning of a data set into subsets so that the data in each subset share some common trait. The measure for a common trait is defined before the clustering is performed and is often a distance metric defining the relative similarity between the two objects. Data clustering is a common technique for statistical data analysis, and has applications to many fields, including machine learning, data mining, pattern recognition, image analysis, and bioinformatics.

Clustering gene expression data helps in identifying genes of similar function. These co-expressed genes with poorly characterised or novel genes may provide a simple means of gaining insight to the functions of many genes for which information is not available currently (11). Co-regulated families of genes cluster together, as was demonstrated by the clustering of ribosomal genes as a group (12). Clustering is also used to identify the grouping patterns of specimens and has been widely used in studying the heterogeneity of cancer. Clinical breast cancers cluster as distinct groups based on their gene expression profiles and can be correlated with clinical outcomes (13, 14).

Primarily, most clustering techniques use a distance metric to define the similarity or difference between the two objects. Some of the most common distance metric used is Euclidean distance, Manhattan distance, and Correlation distance. Euclidean distance is the distance between two points that would be measured with a simple ruler, and can also be calculated by repeated application of the Pythagorean Theorem. Thus, the distance measure would be

$$\text{Distance} = \sqrt{\left(\sum (X_i - Y_i)^2\right)}$$

Manhattan distance is the distance between two points expressed as the sum of the absolute differences of their coordinates. Therefore, the distance between point P_1 with coordinates (x_1, y_1) and the point P_2 at (x_2, y_2) would be $|x_1 - x_2| + |y_1 - y_2|$.

Correlation distance measures the similarity between two points expressed as the correlation between the two objects. Often, Pearson correlation measure is taken as distance measure for most of the microarray data clustering. Correlation measure ranges from −1 to +1. Positive values indicate a positive correlation (i.e. increase in value of one corresponds to increase in the value of the other). Negative values indicate a negative correlation (i.e. increase in value of one corresponds to decrease in value of the other and vice versa). A correlation value of 0 indicates no relation between the two values.

3.4.1. Hierarchical Clustering

Hierarchical clustering is a technique to generate a hierarchy among objects based on their similarity or differences. The similarity or difference is measured based on the distance criteria explained above. Hierarchical clustering may be constructed using an agglomerative or divisive approach. The representation of this hierarchy is a tree also known as dendrogram, with individual elements at one end and a single cluster containing every element at the other (Fig. 5). Agglomerative algorithms begin at the leaves of the tree, whereas divisive algorithms begin at the root. Agglomerative clustering can be single linkage clustering, complete linkage clustering, or average linkage clustering.

Single linkage clustering: The distance between groups is defined as the distance between the closest pair of objects, and only pairs consisting of one object from each group are considered (Fig. 6).

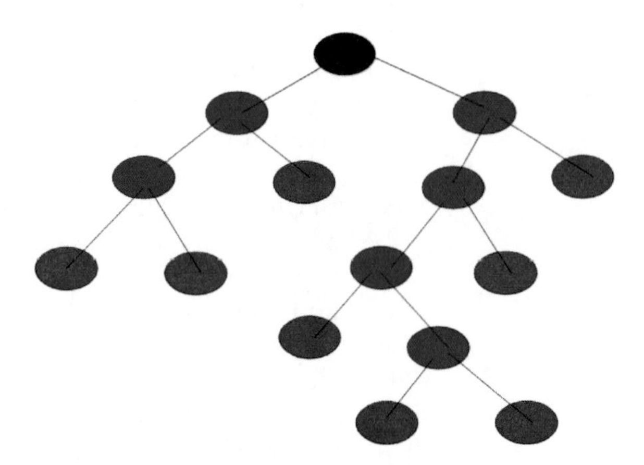

Fig. 5. An example of a tree or dendrogram. The leaves are shown in *red* and the nodes are shown in *blue*. A leaf reflects the entity and a node reflects the relationship between two entities, one entity and one node, or between two nodes.

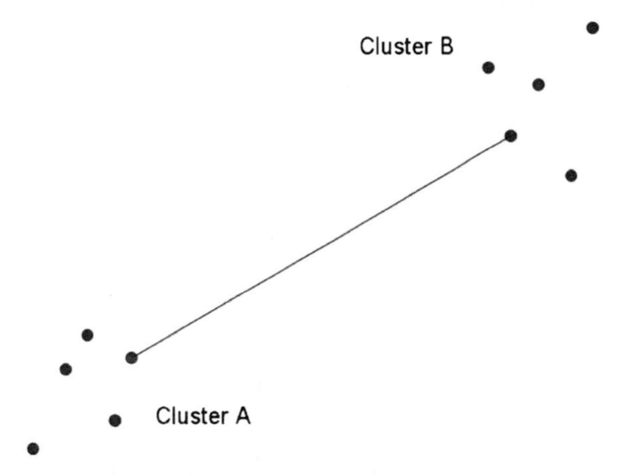

Fig. 6. Single linkage clustering. The closest element in the cluster is used to calculate the reference distance among the two clusters.

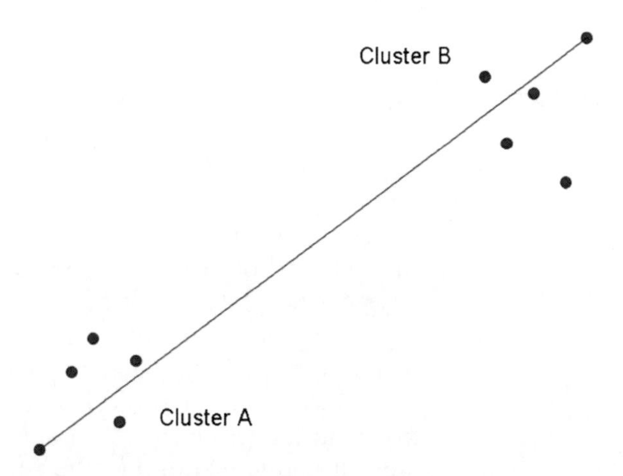

Fig. 7. Complete linkage clustering. The most distant element in the cluster is used to calculate the reference distance among the two clusters.

Complete linkage clustering: The complete linkage, also called farthest neighbour clustering method, is the opposite of single linkage. The distance between groups is defined as the distance between the most distant pair of objects, one from each group (Fig. 7).

Average linkage clustering: Distance between two clusters is defined as the average of distances between all pairs of objects, where each pair is made up of one object from each group (Fig. 8).

Hierarchical clustering has been extensively used to identify relationship among genes and samples. Hierarchical clustering using multiple markers can group breast cancers into various classes with clinical relevance and is superior to individual prognostic markers (15). Hierarchical clustering has been widely used in studying the subgroups in breast cancer (13, 16–18).

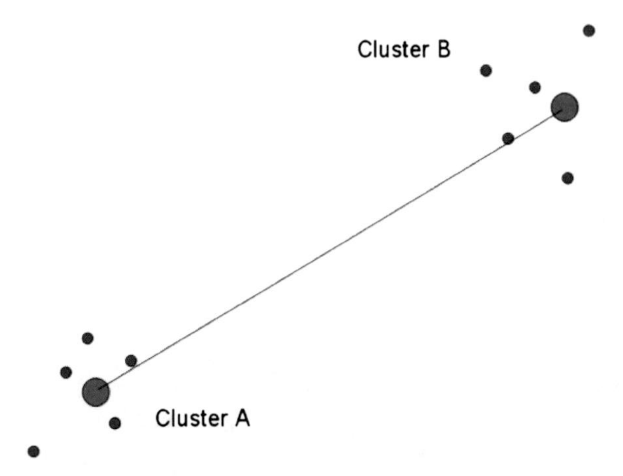

Fig. 8. Average linkage clustering. The average of the element in the cluster is used to calculate the reference distance between the two clusters. The *green* is the average or centroid of the cluster.

3.4.2. k-Means Clustering

The *k*-means algorithm is an algorithm to cluster objects into *k* partitions using the similarity between the objects. *k* is the number of partitions/clusters and is provided by the user. The algorithm starts by partitioning the input points into *k* initial sets randomly or by using some heuristic approaches. It then calculates the centroid (mean point) of each set. Thereafter, it constructs a new partition by associating each object with the closest centroid. The centroids are then recalculated for the new clusters, and the process repeated by alternate application of these two steps until convergence, which is obtained when the objects no longer switch clusters or the centroids no longer change. *k*-means is one of the most commonly used clustering methods and has a wide application in microarray studies (19).

3.5. Principal Component Analysis

Principal component analysis (PCA) is a method to reduce multidimensional data sets to lower dimensions for easier analysis and visualisation. PCA is mathematically defined as an orthogonal linear transformation that transforms the data to a new coordinate system such that the greatest variance by any projection of the data comes to lie on the first coordinate (called the first principal component), the second greatest variance on the second coordinate, and so on.

PCA can be used for dimensionality reduction in a data set by retaining those features of the data set that contribute most to its variance, by keeping lower-order principal components and ignoring higher-order ones. Lower-order components contain the most important essence of the data and higher-order components contain the least important essence of the data. However, this may not be the case with all types of data sets. PCA is used in microarray experiments to identify the most significant patterns in the data.

Raychaudhuri et al. (20), working with yeast sporulation data, concluded that much of the observed variability in the experiment was captured by the first two components corresponding to overall induction level and change in induction level over time.

3.6. Gene Ontology and Pathways Analysis

Microarray experiments result in large number of genes which are differentially expressed across various conditions. The tough part lays in interpretation of the biology of the genes and the conditions. One way is to map the differentially expressed genes to ontology, and known pathways. Gene ontology (GO) is a controlled vocabulary that defines the function of genes and proteins (21). The GO is classified as biological process, cellular component, and molecular function. Under each of these subclassifications, there are number of categories in a highly nested order (Fig. 9).

The genes are associated with functional terms. The aim of the analysis is to find enriched terms among the differentially expressed genes, or a set of co-regulated genes identified by clustering or other techniques. Many algorithms have been proposed to identify the enriched GO term (22). These algorithms compute the number of genes for a particular GO term and the probability of them coming by change and uses mathematical models to identify whether the genes of a particular function have been over- or under-represented.

Same concept is extended to pathways analysis, and the algorithms pull out those pathways which have differentially expressed genes above the threshold. The threshold is calculated as a probability of genes to appear in the individual pathways by chance.

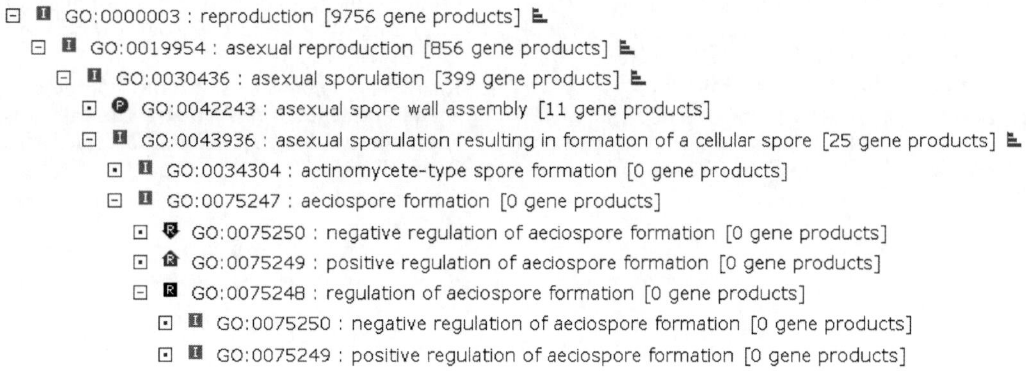

GO:0000003 : reproduction [9756 gene products]
GO:0019954 : asexual reproduction [856 gene products]
GO:0030436 : asexual sporulation [399 gene products]
GO:0042243 : asexual spore wall assembly [11 gene products]
GO:0043936 : asexual sporulation resulting in formation of a cellular spore [25 gene products]
GO:0034304 : actinomycete-type spore formation [0 gene products]
GO:0075247 : aeciospore formation [0 gene products]
GO:0075250 : negative regulation of aeciospore formation [0 gene products]
GO:0075249 : positive regulation of aeciospore formation [0 gene products]
GO:0075248 : regulation of aeciospore formation [0 gene products]
GO:0075250 : negative regulation of aeciospore formation [0 gene products]
GO:0075249 : positive regulation of aeciospore formation [0 gene products]

Fig. 9. Hierarchal architecture of gene ontology terms.

References

1. Derisi, J. (2001) Overview of nucleic acid arrays. *Curr Protoc Mol Biol.* Chapter 22, Unit 22.1.
2. Kerr, M.K, Churchill, G.A. (2007) Statistical design and the analysis of gene expression microarray data. *Genet Res.* **89**, 509–14.
3. Vinciotti, V., Khanin, R., D'Alimonte, D., *et al.* An experimental evaluation of a loop versus a reference design for two-channel microarrays. *Bioinformatics.* **21**, 492–501.
4. Lee, M.L., Kuo, F.C., Whitmore, G.A., Sklar, J. (2000) Importance of replication in microarray gene expression studies: statistical methods and evidence from repetitive cDNA hybridizations. *Proc Natl Acad Sci USA.* **97**, 9834–9.
5. Pavlidis, P., Li, Q., Noble, W.S. (2003) The effect of replication on gene expression microarray experiments. *Bioinformatics.* **19**, 1620–7.
6. Smyth, G.K., Speed, T. (2003) Normalization of cDNA microarray data. *Methods.* **31**, 265–73.
7. Pepper, S.D., Saunders, E.K., Edwards, L.E., Wilson, C.L., Miller, C.J. (2007) The utility of MAS5 expression summary and detection call algorithms. *BMC Bioinformatics.* **8**, 273.
8. Irizarry, R.A., Bolstad, B.M., Collin, F., Cope, L.M., Hobbs, B., Speed, T.P. (2003) Summaries of Affymetrix GeneChip probe level data. *Nucleic Acids Res.* **31**, e15.
9. Li, C., Hung, Wong, W. (2001) Model-based analysis of oligonucleotide arrays: model validation, design issues and standard error application. *Genome Biol.* **2**, RESEARCH0032.
10. Reiner, A., Yekutieli, D., Benjamini, Y. (2003) Identifying differentially expressed genes using false discovery rate controlling procedures. *Bioinformatics.* **19**, 368–75.
11. Eisen, M.B., Spellman, P.T., Brown, P.O., Botstein, D. (1998) Cluster analysis and display of genome-wide expression patterns. *Proc Natl Acad Sci USA.* **95**, 14863–8.
12. Alon, U., Barkai, N., Notterman, D.A., *et al.* (1999) Broad patterns of gene expression revealed by clustering analysis of tumor and normal colon tissues probed by oligonucleotide arrays. *Proc Natl Acad Sci USA.* **96**, 6745–50.
13. Sorlie, T., Perou, C.M., Tibshirani, R., *et al.* (2001) Gene expression patterns of breast carcinomas distinguish tumor subclasses with clinical implications. *Proc Natl Acad Sci USA.* **98**, 10869–74.
14. Sorlie, T., Tibshirani, R., Parker, J., *et al.* (2003) Repeated observation of breast tumor subtypes in independent gene expression data sets. *Proc Natl Acad Sci USA.* **100**, 8418–23.
15. Makretsov, N.A., Huntsman, D.G., Nielsen, T.O., *et al.* (2004) Hierarchical clustering analysis of tissue microarray immunostaining data identifies prognostically significant groups of breast carcinoma. *Clin Cancer Res.* **10**, 6143–51.
16. Charafe-Jauffret, E., Ginestier, C., Monville, F., *et al.* (2006) Gene expression profiling of breast cell lines identifies potential new basal markers. *Oncogene.* **25**, 2273–84.
17. Weigelt, B., Hu, Z., He, X., *et al.* (2005) Molecular portraits and 70-gene prognosis signature are preserved throughout the metastatic process of breast cancer. *Cancer Res.* **65**, 9155–8.
18. Hu, Z., Fan, C., Oh, D.S., *et al.* (2006) The molecular portraits of breast tumors are conserved across microarray platforms. *BMC Genomics.* 7, 96.
19. Do, J.H., Choi, D.K. (2008) Clustering approaches to identifying gene expression patterns from DNA microarray data. *Mol Cells.* **25**, 279–88.
20. Raychaudhuri, S., Stuart, J.M., Altman, R.B. (2000) Principal components analysis to summarize microarray experiments: application to sporulation time series. *Pac Symp Biocomput.* 455–66.
21. Ashburner, M., Ball, C.A., Blake, J.A., *et al.* (2000) Gene ontology: tool for the unification of biology. The Gene Ontology Consortium. *Nat Genet.* **25**, 25–9.
22. Khatri, P., Draghici, S. (2005) Ontological analysis of gene expression data: current tools, limitations, and open problems. *Bioinformatics.* **21**, 3587–95.

Chapter 4

Software and Tools for Microarray Data Analysis

Jai Prakash Mehta and Sweta Rani

Abstract

A typical microarray experiment results in series of images, depending on the experimental design and number of samples. Software analyses the images to obtain the intensity at each spot and quantify the expression for each transcript. This is followed by normalization, and then various data analysis techniques are applied on the data. The whole analysis pipeline requires a large number of software to accurately handle the massive amount of data. Fortunately, there are large number of freely available and commercial software to churn the massive amount of data to manageable sets of differentially expressed genes, functions, and pathways. This chapter describes the software and tools which can be used to analyze the gene expression data right from the image analysis to gene list, ontology, and pathways.

Key words: Microarray, Gene expression, Normalization, Clustering, Classification, Gene ontology, Pathways

1. Introduction

A microarray experiment generates a large volume of data, originally as image files which are later analyzed into meaningful results. The whole process involves a number of steps and a number of software and tools to obtain dependable expression values for each gene. With the information in hand, the next obvious question is to get the biology out of the data. A lot of software has been developed to answer this question. This section describes in brief important tools and software for microarray data analysis.

Affymetrix gene chip platform is the most popular commercially available platforms for gene expression study. It is estimated that Affymetrix share of the global commercial microarray market is above 75% (1). This had led to the development of large number of software and tools for gene chip data analysis; few by Affymetrix and many more developed academically. Besides that, there are lots

Lorraine O'Driscoll (ed.), *Gene Expression Profiling: Methods and Protocols*, Methods in Molecular Biology, vol. 784, DOI 10.1007/978-1-61779-289-2_4,

of commercial software for Affymetrix gene chip data analysis. Here, we describe few of the software specific for Affymetrix data analysis.

2. Affymetrix Pipeline Software

Affymetrix pipeline software comes with the hardware and controls the hardware and the fluidics. Additionally, they perform image acquisition, expression quantification, quality control, data normalization, and few other statistical analyses. These software include GeneChip® Operating Software (GCOS), Affymetrix® GeneChip® Command Console® Software (AGCC), and Affymetrix® Expression Console™ Software and are described below.

2.1. GeneChip® Operating Software

Affymetrix hardware comes with the software to manage the hardware, perform image analysis, expression quantification, and data normalization. The software controls the Affymetrix fluidics and performs image acquisition, and quantification of expression values. Besides that, it also estimates the quality control parameters and performs normalization.

2.2. Affymetrix® GeneChip® Command Console® Software

GCOS has been superseded by AGCC. AGCC controls the fluidics station and provides easy data management and supports easy integration with Affymetrix® Expression Console™ Software and other third party software. AGCC supports GeneTitan™ Instrument, Fluidics Station 450, and GeneChip® Scanner 3000.

2.3. Affymetrix® Expression Console™ Software

Affymetrix® Expression Console™ Software integrates with GCOS and AGCC. The software performs the probe set summarization, quantification, and normalization of gene chip expression arrays. The software is equipped with Microarray Suite 5.0 (MAS5) normalization algorithm, Probe Logarithmic Intensity Error Estimation (PLIER) normalization, and Robust Multichip Analysis (RMA) normalization.

3. Free Software

These software are available free for the academic users.

3.1. RMA Express

RMA express software (2) implements RMA normalization algorithm (3) and is compatible with Windows and Linux operating system. The software has a graphic user interface with options to select the Affymetrix raw data files and perform normalizing and quantifying of the expression values. The latest version included many of the QC parameters, including visualization of the chip, box plots, and density plots. The software is free for academic use and can be downloaded from http://rmaexpress.bmbolstad.com. RMA normalization can also be performed using R (http://www.r-project.org) and Bioconductor (http://www.bioconductor.org).

3.2. dCHIP

DNA-Chip Analyzer (dCHIP) was originally developed by Cheng Li and Wing Hung Wong and implements a model-based expression analysis for Affymetrix gene expression arrays (4). The software has a capability to process Affymetrix raw data (dat and cel files) and processed data (quantified expression values as a tab delimited file). With time, the software has grown to perform large number of analysis and have also included modules for other types of Affymetrix arrays, e.g. SNP array, exon array, tiling arrays, etc. The salient features of this software are summarized below.

1. *Normalization and quality control* (QC). The software processes raw data, normalize them, and quantifies as a single expression values for each transcript. Additionally, a file detailing the QC parameters is also generated. This includes Un-normalized Median Intensity, Percent of Present call Array outlier, and Single outlier along with the warning, if any. The individual probes can be simulated as an image to get a deeper understanding of any underlying problems. An example of a good and a bad chip is shown in Fig. 1.
2. *Hierarchical clustering*. The hierarchal clustering module in dCHIP has been extensively developed to accommodate chromosome, gene ontology (GO), and pathways information. The module intuitively identifies clusters of samples with

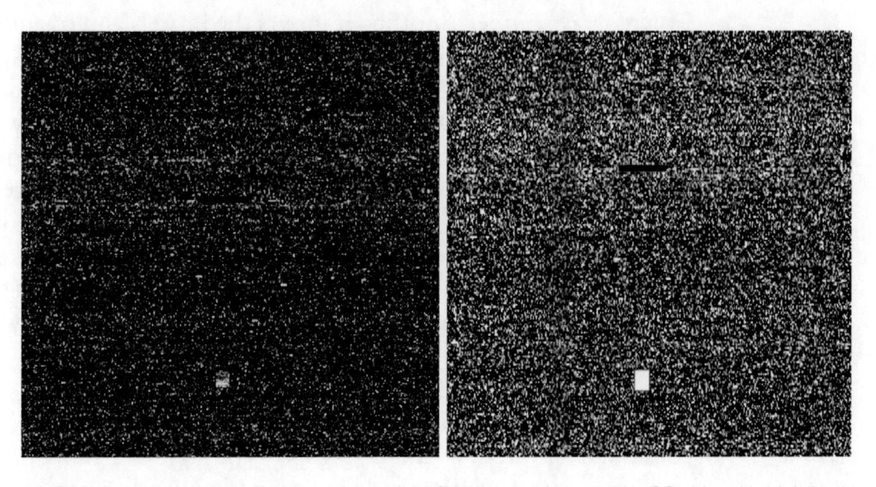

Fig. 1. The *left figure* represents a good chip and the *right figure* represents a poor QC chip with high background. The images were generated using dCHIP.

enriched clinical parameters and also identifies clusters of genes with genes enriched for particular gene ontology, pathway, and chromosome. Options are provided to rearrange the nodes of tree and export a section of the tree as image file or gene lists.

3. *Compare samples.* Most of the gene expression studies are designed to compare set of samples to identify differentially expressed (DE) genes. Fold change and *t*-test had been the gold standards in identifying DE genes. dCHIP "Compare samples" compares two sets of samples for DE genes using the user-defined criteria which can be fold change, difference in expression, *p*-value, or combination of these. The option to upload the specimen information file or clinical information file makes the comparison very easy and less prone to error. The analysis includes the number of differentially expressed genes and also accesses the false discovery rate (FDR).

Additionally, there are many other exciting features for the data analysis, including mapping on chromosome, gene ontology, and pathways analysis and various exporting options. The software is free for academic use and can be downloaded from http://www.dchip.org.

3.3. TM4

TM4 suite of tools includes four major applications, Microarray Data Manager (MADAM), TIGR Spotfinder, Microarray Data Analysis System (MIDAS), and Multiexperiment Viewer (MeV) (5). This software is free for academic use and can be downloaded from http://www.tm4.org/. This software was primarily designed for cDNA chips, but can be adapted for Affymetrix arrays. Together, they make up a complete pipeline for microarray data analysis. The details of this software are described below.

3.3.1. Spotfinder

For custom-designed cDNA array, the chips are scanned using a microarray scanner. Spotfinder inputs the images and quantifies the intensity as a measurable expression value.

Spotfinder is capable of reading both 8- and 16-bit images generated by most of the microarray scanners. The software has the ability to perform automated or semi-automated grid quantification. Following this, spot intensity is calculated as integral of non-saturated pixels followed by calculation of the background intensity. It then outputs the mean, median, and background intensity of each spot along with the quality control flag. The software is free to use and can be downloaded from http://www.tm4.org/spotfinder.html.

3.3.2. Microarray Data Analysis System

MIDAS is a part of TM4 group of software. The software is designed to process the raw expression values from Spotfinder, perform normalization, and data adjustment techniques along with various filtration options. Important normalization techniques available include global normalization, LOWESS normalization, iterative linear regression normalization, and standard deviation regularization. The quality check module includes slice analysis (z-score filtering) and flip dye consistency checking. The data filtration module includes MAANOVA, t-test, and significance analysis of microarrays (SAM).

3.3.3. Multiexperiment Viewer

MeV is comprehensive software designed to perform large number of analysis. This software comes with exhaustive list of methods and algorithms which can be applied to the gene expression data. Some of the important analyses supported by MeV are many types of data transformation and normalization; finding significant genes, such as t-test, SAM, non-parametric tests, ANOVA, etc.; clustering techniques, such as hierarchical clustering, self-organizing maps, k-means clustering, and many more; classification techniques, such as k-nearest neighbour, discriminate analysis, and support vector machines (SVMs); data reduction techniques, such as principal component analysis, correspondence analysis, relevance network, etc. The software is free for academic use and can be downloaded from http://www.tm4.org/mev.html.

3.4. SNOMAD

SNOMAD is a web-based tool and has various normalization options for two-channel and single-channel experiments (6). The data need to be uploaded in appropriate format and the choice of normalization techniques selected. The options include background subtraction, global mean normalization, local mean normalization across a microarray surface, log transformation, calculation of mean log intensity and log ratio local mean normalization across element signal intensity, and local variance correction across element signal intensity. The software is free to use and can be accessed from http://pevsnerlab.kennedykrieger.org/snomadinput.html.

3.5. Significance Analysis of Microarrays

SAM is an excel add-on tool to identify DE genes (7). The software can analyze both two-channel and Affymetrix expression values, has basic normalization options, and can deal with missing data. The different experimental designs supported by SAM include:

1. *Two-class experiments.* Such experiments have two groups of samples and the aim is to identify genes up- or downregulated among the two groups.
2. *Two-class experiments.* Such experiments are similar to above; however, the experimental blocks are defined.
3. *Paired.* Such experiments have matching pair of samples and up- and downregulation are defined as differences among the matched pairs rather than the holistic differences among the two groups.

4. *Multiclass*. Such experiments have more than two-class and the aim is to find DE genes.
5. *Time series experiment*. The gene expression measures are available at different time points, and the aim is to find genes with gene expression patterns correlating the time.
6. *Two-class time series*. Such experiments have two groups of samples and each group have time points, and the aim of the experiment is to find DE genes across the time points.
7. *Two-class paired time series*. Such experiments are similar to above; however, the samples in two groups are paired.
8. *Survival analysis*. Such experiments have censored survival times of patients, and the aim is to find genes which correlate with survival.

The output of SAM contains the gene-specific information along with SAM score, numerator, denominator, q-value, and local FDR. "SAM score" is the value of t-statistics or Wilcoxon statistics depending on the initial option selected by the user. Higher is this value, more significant is the difference. The "Numerator" is the numerator of the test statistics and "Denominator" is the denominator of the test statistics. "q-value" is the lowest FDR at which the gene is called significant (8). Local FDR is calculated for those genes where SAM score falls in a window close to the gene under consideration.

To obtain list of significant genes, the software has a delta slider or can be manually entered. By altering the delta score, the number of significant genes and the FDR is listed and the user can select the list with the desired FDR. The software also has option to add fold change as additional filtration criteria. The software can be obtained from http://www-stat.stanford.edu/~tibs/SAM/.

3.6. Genesis

Genesis (9) is a comprehensive software with the ability to process and analyze large microarray datasets into meaningful results. The software has modules for cDNA microarray data transformation and normalization and large number of clustering options, e.g. hierarchical clustering, k-means clustering (Fig. 2), self-organizing maps, etc. Additionally, it has features like SVMs and principal component analysis. The software is easy to use and is quite robust to handle large datasets. The software is free to use for academic purpose and can be downloaded from http://genome.tugraz.at/genesisclient/genesisclient_description.shtml.

3.7. Gene Expression Model Selector

Gene expression model selector (GEMS) is a software to develop a prediction classifier based on the known outcomes (10). The software implements support vector machines and have support for multiclass classification. This software has been designed to input

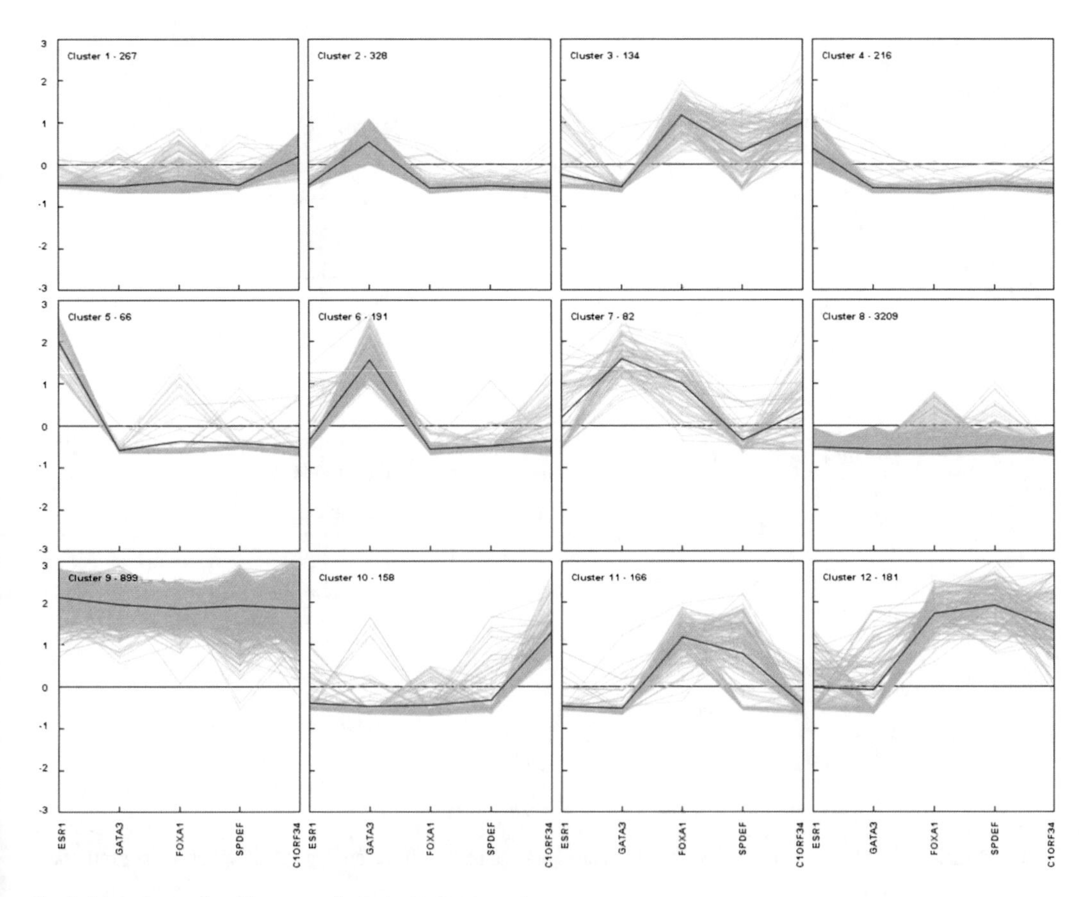

Fig. 2. Analysis results of *k*-means clustering using genesis.

gene expression data from cancer patients and develop a model on the known data to predict the outcome of unknown data. The developers of this software analyzed 11 different datasets of subclass of various types of cancer and demonstrated that SVM outperformed other classifiers when it comes to multiclass classification. The software has a number of options to optimize the model to obtain the most accurate results. Polynomial kernel, including linear kernel and Gaussian kernel, has been implemented in the software. The user can perform N-fold cross-validation of the data to identify the prediction accuracy of the system. Such validation is very important as it ensures that the model has not included the test data and the accuracy is on the test data rather than on the training data. Since in most such experiments, the number of samples is less, the obvious method of choice is N-fold- or leave-one-out cross-validation. The software is very easy to operate and the input file requires the expression values along with the subclass and the model is developed on that. Basic normalization and variable selection modules are also provided in the software. The software can be downloaded from http://www.gems-system.org/.

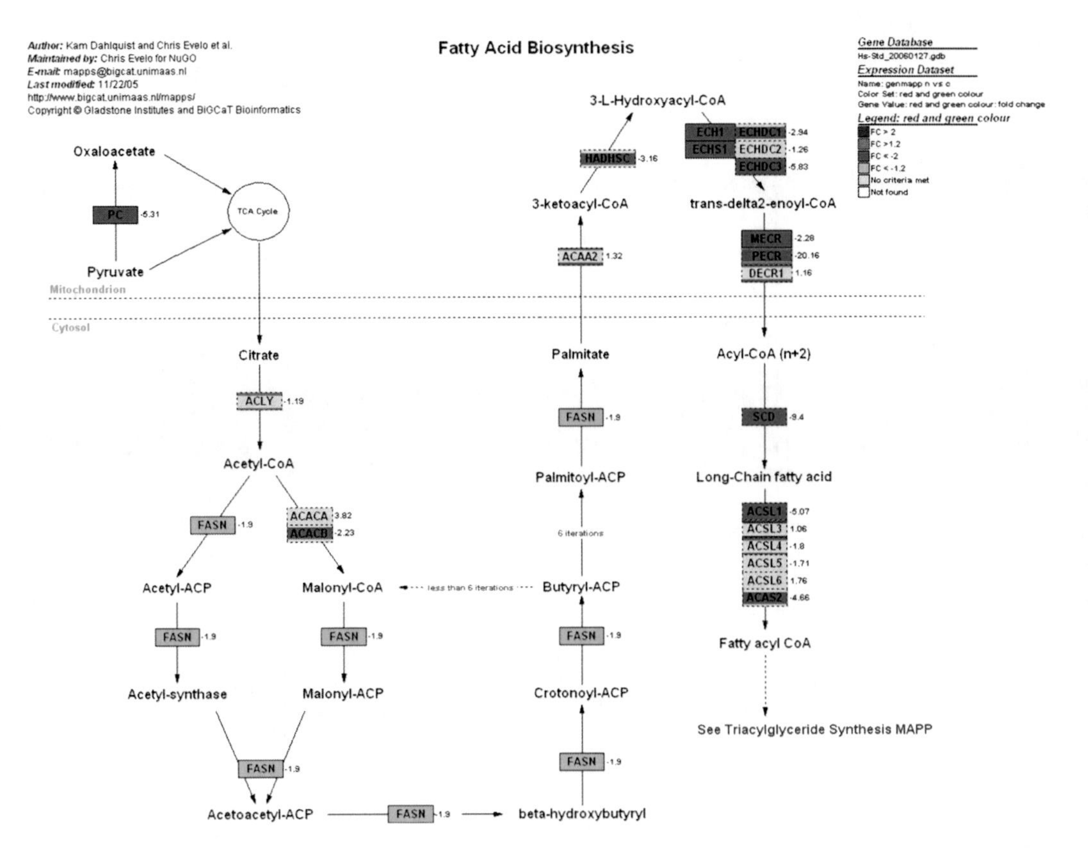

Fig. 3. Fatty acid biosynthesis pathway overlaid with gene expression data. The colour indicated over or under expression.

3.8. GenMAPP

GenMAPP (11) is a computer application designed to visualize gene expression and other genomic data on maps representing biological pathways and groupings of genes. It overlays gene expression data on the pathways, incorporating colour-code according to user-defined parameters (Fig. 3). Additionally, the MAPPFinder module identifies significant gene ontologies and pathways affected by the submitted gene lists.

3.8.1. MAPPFinder

MAPPFinder (12) is an accessory programme that works with GenMAPP and the annotations from the Gene Ontology Consortium to identify significant GO and MAPPs. The calculations made by MAPPFinder (Fig. 4) are intended to give an idea of the relative amount of genes meeting the criterion that are present in each GO term or local MAPP.

Genes meeting the criterion. The number of distinct genes that met the user-defined criterion in the expression dataset. This may also be referred to as "genes changed."

Genes measured. The number of distinct genes in the submitted expression dataset that were found to link to this GO term or MAPP.

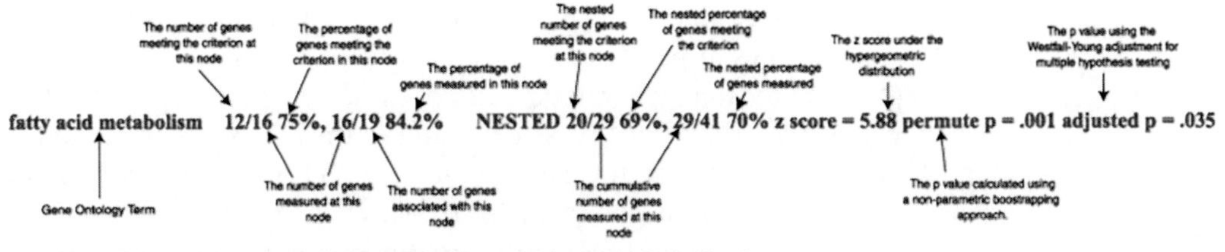

Fig. 4. Calculations made by GenMAPP to calculate the significance of individual GO and MAPPs (Obtained from GenMAPP Web site).

$$zscore = \frac{(r - n\frac{R}{N})}{\sqrt{n(\frac{R}{N})(1-\frac{R}{N})(1-\frac{n-1}{N-1})}}$$

Fig. 5. *Z*-score calculation (Obtained from GenMAPP Web site).

Genes associated with this GO term or MAPP: The number of genes assigned to this GO term or on this MAPP. Also referred to as the number of "Genes in GO" for a specific term.

% Genes meeting the criterion. Genes meeting the criterion/genes measured × 100.

% Genes measured. Genes measured/genes associated × 100.

Nested numbers. The same five calculations are repeated, but as nested numbers.

Z score. The standard statistical test under the hypergeometric distribution (Fig. 5). where N is the total number of genes measured, R is the total number of genes meeting the criterion, n is the total number of genes in this specific MAPP, and r is the number of genes meeting the criterion in this specific MAPP. A positive Z score indicates that there are more genes meeting the criterion in a GO term/MAPP than would be expected by random chance. A negative Z score indicates that there are fewer genes meeting the criterion than would be expected by random chance.

Permute P and adjusted P. p-value is calculated based on the Z score.

GenMAPP is free for academic use and can be downloaded from www.genmapp.org.

3.9. Onto-Express and Pathway-Express

Onto-Express (13, 14) is a tool that automatically translates DE gene transcripts from microarray experiments into functional profiles characterizing the impact of the condition studied. The software constructs functional profiles (using GO terms) for the categories: biochemical function, biological process, cellular role, cellular component, molecular function, and chromosome location and calculates significance values for each category.

Pathway-Express (15, 16) works on the provided list of genes and the system performs a search and builds a list of all associated pathways. After generating a list of pathways for the input list of genes from the Onto-Tools database, Pathway-Express calculates a perturbation factor for each input gene. This perturbation factor takes into account the normalized fold change of the gene and the number and amount of perturbation of genes downstream from it. This gene perturbation factor reflects the relative importance of each DE gene. The impact factor of the entire pathway includes a probabilistic term that takes into consideration the proportion of DE genes on the pathway and gene perturbation factors of all genes in the pathway. The impact factors of all pathways are used to rank the pathways before presenting them to the user.

Both Onto-Express and Pathway-Express are free for academic use and can be downloaded from http://vortex.cs.wayne.edu/projects.htm.

3.10. Database for Annotation, Visualization and Integrated Discovery and Expression Analysis Systematic Explorer

Database for Annotation, Visualization and Integrated Discovery (DAVID) (17) is an online tool for annotation and functional analysis. Expression Analysis Systematic Explorer (EASE) (18) is a downloadable version of DAVID with few added features. They comprise four distinct modules which are described below.

Functional annotation. This module performs gene set enrichment analysis, clustering genes based on their function annotation, identifying, and mapping gene sets to pathways, such as BioCarta and KEGG, gene-disease association, homologue match, ID translation, and literature match.

Gene functional classification. This module reduces large list of genes into functionally related groups thereby making it perfect for high-throughput technologies.

Gene ID conversion. This module converts any identifier to specified identifier using its extensive mapping database. This becomes very valuable tool when comparing gene lists from two different platforms or experiments.

Gene name batch viewer. This module displays gene names for the list provided and searches for functionally related genes in the list or not in the list.

The tool is free to use and can be accessed online at http://david.abcc.ncifcrf.gov.

3.11. Venny

Venny is an online tool to compare gene list obtained from different comparisons or experiment. Venn diagram is quite often used to compare and depict the results graphically. Venn diagram is ideally suited for comparing up to four gene lists. The common area shows the number of genes common to two or more gene lists (Fig. 6). The tool is free to use for non-profit use and can be accessed from http:// bioinfogp.cnb.csic.es/tools/venny/index.html.

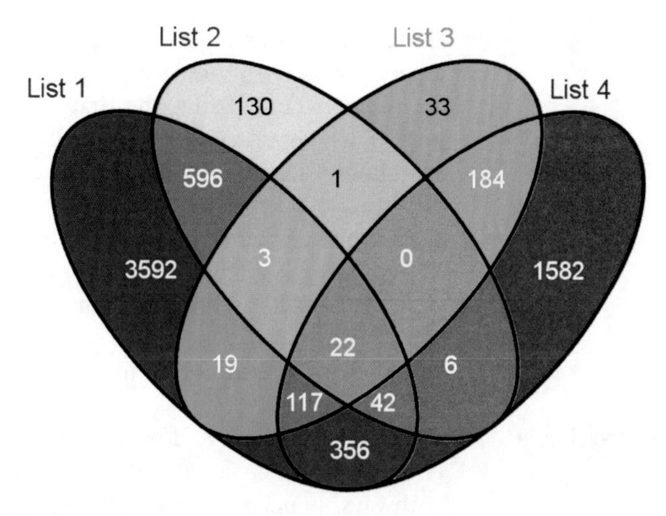

Fig. 6. The figure shows common genes among two or more gene lists. The image was created using Venny.

4. Commercial Software

There are large numbers of commercial software for microarray data analysis. Mostly, they are comprehensive software and include large number of features with extensive visualization and graphics. The extensive support provided by the vendors makes it attractive for the labs where they cannot afford a full-time bioinformatician to analyze gene expression data. Some of these software are listed below.

4.1. Genespring

Genespring from Agilent (earlier Silicon Genetics) was the first of the commercial software available for microarray data analysis. It comes with large number of tools for microarray data analysis.

4.2. Partek

Genomics Suite from Partek is a group of software modules for analyzing gene expression data and various other types of data, such as Alternative Splicing, Copy Number, SNP Association, and Next Generation Sequencing.

4.3. Genowiz

Genowiz from Ocimumbio Solutions is a comprehensive software for gene expression analysis compatible to different data formats and have large number of features for various analysis.

4.4. GenomatixSuite

GenomatixSuite from Genomatix is software for mining biological relevance of the genes of interest obtained from microarray analysis. The suite includes ElDorado, Gene2Promoter, GEMS Launcher, MatInspector, MatBase, and BiblioSphere. The software mines the available literature and build interaction maps among the genes of interest. So if gene A is co-cited in literature with gene B, and gene B is co-cited with gene C, the software will construct an interaction map showing that there is an interaction between A and C which normally could not be found in normal literature searches. Additionally, the software is capable to mine promoter and transcriptional factor region for genes of interest and builds transcriptional factor models which may regulate a bunch of genes. This module is ideally suited to study genes with similar expression pattern to identify transcriptional models, which may be regulating the expression of genes. The software is expensive and is priced at 1,300 euros/month for single-user licence.

4.5. Pathway Studio

Pathway Studio from Ariadne Genomics (http://www.ariadnegenomics.com) is a product aimed at the visualization and analysis of biological pathways, gene regulation networks, and protein interaction maps from the list of genes from a microarray experiment. It comes with a comprehensive database that gives a snapshot of all information available in PubMed, with the focus on pathways and cell-signalling networks. MedScan is a part of Pathway Studio and helps in mining current literature and building pathways related to the topic of interest, such as disease, phenotype, or physiological process. The software downloads all PubMed abstracts, extract relations from them, and add them to the selected pathway.

References

1. Rubenstein, K. (2003) Commercial aspects of microarray technology. *BioTechniques.* Suppl: 52–4.
2. Bolstad, B.M., Irizarry, R.A., Astrand, M., Speed, T.P. (2003) A comparison of normalization methods for high density oligonucleotide array data based on variance and bias. *Bioinformatics.* **19**, 185–93.
3. Irizarry, R.A., Bolstad, B.M., Collin, F., Cope, L.M., Hobbs, B., Speed, T.P. (2003) Summaries of Affymetrix GeneChip probe level data. *Nucleic Acids Res.* **31**, e15.
4. Li, C., Hung, Wong. W. (2001) Model-based analysis of oligonucleotide arrays: model validation, design issues and standard error application. *Genome Biol.* **2**, RESEARCH0032.
5. Saeed, A.I., Sharov, V., White, J., *et al.* (2003) TM4: a free, open-source system for microarray data management and analysis. *BioTechniques.* **34**, 374–8.
6. Colantuoni, C., Henry, G., Zeger, S., Pevsner, J. (2002) SNOMAD (Standardization and NOrmalization of MicroArray Data): web-accessible gene expression data analysis. *Bioinformatics.* **18**, 1540–1.
7. Tusher, V.G., Tibshirani, R., Chu, G. (2001) Significance analysis of microarrays applied to the ionizing radiation response. *Proc Natl Acad Sci USA.* **98**, 5116–21.
8. Efron, B., Tibshirani, R. (2002) Empirical bayes methods and false discovery rates for microarrays. *Genet Epidemiol.* **23**, 70–86.
9. Sturn, A., Quackenbush, J., Trajanoski, Z. (2002) Genesis: cluster analysis of microarray data. *Bioinformatics.* **18**, 207–8.
10. Statnikov, A., Aliferis, C.F., Tsamardinos, I., Hardin, D., Levy, S. (2005) A comprehensive evaluation of multicategory classification methods for microarray gene expression cancer diagnosis. *Bioinformatics.* **21**, 631–43.
11. Dahlquist, K.D., Salomonis, N., Vranizan, K., Lawlor, S.C., Conklin, B.R. (2002) GenMAPP, a new tool for viewing and analyzing microarray data on biological pathways. *Nat Genet.* **31**, 19–20.
12. Doniger, S.W., Salomonis, N., Dahlquist, K.D., Vranizan, K., Lawlor, S.C., Conklin, B.R. (2003) MAPPFinder: using Gene Ontology and GenMAPP to create a global gene-expression profile from microarray data. *Genome Biol.* **4**, R7.
13. Draghici, S., Khatri, P., Bhavsar, P., Shah, A., Krawetz, S.A., Tainsky, M.A. (2003) Onto-Tools, the toolkit of the modern biologist:

Onto-Express, Onto-Compare, Onto-Design and Onto-Translate. *Nucleic Acids Res.* **31**, 3775–81.

14. Khatri, P., Draghici, S. (2005) Ontological analysis of gene expression data: current tools, limitations, and open problems. *Bioinformatics.* **21**, 3587–95.
15. Khatri, P., Sellamuthu, S., Malhotra, P., Amin, K., Done, A., Draghici, S. (2005) Recent additions and improvements to the Onto-Tools. *Nucleic Acids Res.* **33**, W762–5.
16. Khatri, P., Voichita, C., Kattan, K., *et al.* (2007) Onto-Tools: new additions and improvements in 2006. *Nucleic Acids Res.* **35**, W206–11.
17. Dennis, G., Jr, Sherman, B.T., Hosack, D.A., *et al.* (2003) DAVID: Database for Annotation, Visualization, and Integrated Discovery. *Genome Biol.* **4**, P3.
18. Hosack, D.A., Dennis, G., Jr, Sherman, B.T., Lane, H.C., Lempicki, R.A. (2003) Identifying biological themes within lists of genes with EASE. *Genome Biol.* **4**, R70.

Chapter 5

Analysis of Gene Expression as Relevant to Cancer Cells and Circulating Tumour Cells

Anne M. Friel, John Crown, and Lorraine O'Driscoll

Abstract

Current literature provides significant evidence to support the concept that there are limited subpopulations of cells within a solid tumour that have increased tumour-initiating potential relative to the total tumour population. Such tumour-initiating cells have been identified in leukaemia and in a variety of solid tumours using different combinations of cell surface markers, suggesting that a tumour-initiating cell heterogeneity exists for each specific tumour. These studies have been extended to endometrial cancer; and herein we present several experimental approaches, both in vitro and in vivo, that can be used to determine whether such populations exist, and if so, to characterize them. These methods are adaptable to the investigation of tumour-initiating cells from other tumour types.

Key words: Endometrial cancer, Tumour-initiating cell, Cancer stem cell, Surface markers, Xenograft model, Bone marrow, Peripheral blood mononuclear cells, Circulating tumour cells

1. Introduction

Stem cells are defined by their capacity for self-renewal, relative quiescence, and infrequent division (1). This allows for the maintenance of an undifferentiated and uncommitted pool of stem cells. Importantly, stem cells are typically "plastic", able to clonally reconstitute a variety of lineage-specific cells. This capacity to self-replicate is maintained under strict homeostatic regulation of cell proliferation, cell differentiation, and cell death.

The concept of the cancer stem cell was initially derived from studies of specific cell populations in human leukaemia. Whether or not the term cancer stem cell is appropriate is still widely debated (2). It is agreed, however, that these tumour-initiating cells have inherent or acquired stem cell-like properties. For continuity, such cells herein are referred to as tumour-initiating cells.

Lorraine O'Driscoll (ed.), *Gene Expression Profiling: Methods and Protocols*, Methods in Molecular Biology, vol. 784, DOI 10.1007/978-1-61779-289-2_5,

Tumour-initiating cells were initially identified in solid tumours based on differential efflux of Hoechst 33342 dye via verapamil-sensitive multidrug resistance transporters (3–6). Additionally, using specific surface markers, tumour-initiating cells have been identified in solid tumours of the breast ($CD44^{+}$, $CD24^{-/low}$, $EpCAM^{+}$, $Lineage^{-}$ (7)), brain ($CD133^{+}$ (8)), colon ($CD133^{+}$ (9)), ovary ($CD44^{+}/CD117^{+}$ (10)), CD133 (11, 12)), and endometrium (13); to name a few. The majority of these studies involved injecting immunocompromised mice with a defined number of cells derived from the bulk of a tumour, or injecting a population of tumour-derived cells with a distinct surface marker profile.

More recent efforts have focused on the identification of the endometrial tumour-initiating cell (13, 14). The continual remodelling of the endometrial lining at menses strongly argues for the presence of a stem/progenitor cell population with regenerative capabilities. An aberrant stem/progenitor cell, within the larger cell population, may result in pre malignant endometrial hyperplasia and/or in endometrial cancer, which is the most common cancer of the female reproductive organs in the USA. The treatment for endometrial cancer is usually surgical followed by an evaluation for metastatic disease. Depending on the extent of disease, patients are advised to receive adjuvant therapy with combination chemotherapy recommended for extra-uterine disease. The efficacy of this chemotherapy is questionable, recurrent disease is common, and there is poor long-term prognosis (15). It is hypothesized that such recurrent tumours arise as a result of the presence of tumour-initiating cells with stem cell-like properties.

The transplantable xenograft mouse model has proven to be a valuable aid in the investigation and characterization of tumour-initiating cells (7–13) and circulating tumour cells (CTCs) in various cancer types (16, 17). This chapter covers the basics on the development of such a model as it pertains to endometrial cancer from the initial processing of the primary carcinoma specimen and/or its surface profiling to injection into NOD/SCID mice, and the analyses of primary carcinoma-derived xenograft tumours. Finally, some additional in vitro and in vivo techniques that cover analyses of CTCs in bone marrow and peripheral blood are included.

2. Materials

2.1. Media and Solutions

1. HBSS/2% FBS/1 mM EDTA. Prepare the 1 mM EDTA solution first and filter sterilize. Add HBSS and FBS to the sterile EDTA solution. Store at 4°C. Ethylenediaminetetraacetic acid (EDTA) (Sigma-Aldrich, St. Louis, MO), Hanks balanced salt solution (HBSS) (Cambrex Corp., East Rutherford, NJ), Foetal bovine serum (FBS) (HyClone, Logan, UT).

2. PBS/2% FBS/1 mM EDTA. As for HBSS/2% FBS/1 mM EDTA, but substitute HBSS with PBS.
3. Collagenase Type II, DNase I, Amphotericin B (Sigma). Store at −20°C.
4. Dulbecco's phosphate-buffered saline (PBS).
5. ACK lysis buffer (Cambrex Corp.).
6. Dulbecco's modified Eagle's medium (DMEM) (Mediatech Inc., Herndon, VA). Store at 4°C. Warm to 37°C before use.
7. 0.4% trypan blue (Mediatech Inc.). Dilute cell suspension 1:5 with trypan blue and count.
8. Paraformaldehyde (PFA) (Electron Microscopy Sciences, Hatfield, PA). Make a 4% solution using sterile PBS.
9. Ficoll-Paque™ PLUS (GE Healthcare Bio-Sciences Corp., Piscataway, NJ).

2.2. Cell Depletion

1. LD column (Miltenyi Biotec Inc., Auburn, CA).
2. MultiStand (Miltenyi Biotec Inc.).
3. Dead Cell Removal Kit (Miltenyi Biotec Inc.).
4. 1× binding buffer: (20× solution supplied in dead cell depletion kit. Dilute in Milli-Q H_2O to make 1× working solution).
5. CD45 Microbeads, human (Miltenyi Biotec Inc.).
6. CD31 Microbeads, human (Miltenyi Biotec Inc.).
7. FcR Blocking Reagent, human (Miltenyi Biotec Inc.).
8. CD133 Microbead kit (Miltenyi Biotec Inc.).
9. Mouse anti-mouse H-$2K^d$ fluorescein isothiocyanate (FITC)-conjugated antibody (BD Biosciences, San Jose, CA).

2.3. Cell Staining

1. Anti-CD133/2 (phycoerythrin) PE-conjugated antibody-human; anti-CD31 FITC-conjugated antibody-human; anti-CD45-FITC-conjugated antibody-human (Miltenyi Biotec Inc.).
2. IgG2b-PE mouse (Miltenyi Biotec Inc.).
3. BD™ CompBead set (BD Biosciences).
4. Anti-CD44-(allophycocyanin) APC-conjugated antibody-human (BD Biosciences).
5. LIVE/DEAD Fixable Dead Cell Stain Kit-Green (Invitrogen, Carlsbad, CA). Working solution: Add 1 μL of LIVE/DEAD to 29 μL of PBS.
6. FcR Blocking Reagent, human, (Miltenyi Biotec Inc.).

2.4. In Vivo Injections

1. NOD/SCID mice (strain NOD.CB17-Prkdcscid/J, Jackson Laboratory, Bar Harbor, ME).
2. Matrigel® (BD Biosciences).

2.5. Retro-Orbital Bleed

1. Avertin (2,2,2 tribromoethanol), T-amyl alcohol (Sigma). Dissolve Avertin in 15.5 mL of T-amyl alcohol. Use a magnetic stirrer. Avertin is light sensitive and hydroscopic. Store the stock solution at 4°C. To make the working solution, add 0.5 mL Avertin to 35.5 mL of sterile normal saline (pre-warmed to 37°C). Ensure that there are no crystal formations as these lead to mouse death by intestinal necrosis if injected. Store the working solution at 4°C in dark. The dose used is 200 mg/kg body weight.
2. Fisherbrand microblood collecting tubes – Heparinized (Fisher Scientific, Pittsburgh, PA).
3. BD Microtainer plastic capillary blood collectors – EDTA coated (Fisher Scientific).
4. RNA *later*™ – (Ambion Inc., Austin, TX).

2.6. Nucleic Acid Isolation

1. RiboPure Blood Kit (Ambion).
2. Wash Solution 2/3 (Supplied in RiboPure Blood Kit). Add 56 mL of 100% ethanol to concentrate.
3. DNeasy Blood & Tissue Kit (Qiagen, Valencia, CA).

3. Methods

3.1. Primary Endometrial Epithelial Cell Isolation

All primary human tissues must be collected in accordance with policies of your Institutional Review Board and/or Ethics Committee. It is recommended that portions of tumour be stored for 4% PFA fixing, OCT mounting, dry frozen, and in 10% DMSO/medium. This will enable future immunohistochemical analyses on FFPE and frozen sections, nucleic acid/protein analyses, and cell line derivation if possible.

1. Transfer the carcinoma tissue to a 100-mm culture dish. Add 1 mL HBSS/2% FBS/1 mM EDTA. Use one blade to secure the tumour piece. With the other blade, start at edge of tumour and chop finely, working along the entire tumour. Add minced tumour to a 50-mL tube. Incubate at 37°C for 1 h with agitation, in HBSS/2% FBS/1 mM EDTA containing 1 mg/mL collagenase Type II/20 mL, 0.025% DNase I, and 1.5 mL amphotericin B. Add 10 mL PBS. Incubate for 10 min at room temperature (see Note 1).
2. Wash pellet with 10 mL PBS.
3. Incubate for 10 min at room temperature (see Note 2).
4. Drain off PBS and add 20 mL DMEM/2% FBS and resuspend the pellet.
5. Transfer contents of the 50-mL tube to a T75 flask.

6. Incubate the T75 flask for 1 h at 37°C, 5% CO_2. Alternate the flask position every 20 min to ensure the flask contents are in contact with three different walls of the flask over the 1-h period.
7. Remove the medium. This contains endometrial epithelial cells, which make up the bulk of the cells in the non-adherent cell population.
8. Assess if cell suspension requires ACK treatment for the removal of red blood cells.
9. Assess the number of non-viable cells, as determined by trypan blue staining, and eliminate from the suspension using a Dead Cell Removal Kit as necessary.

3.2. ACK Lysis-Removal of Red Blood Cells

1. Resuspend cell pellet in 1 mL ACK lysis buffer.
2. Incubate at room temperature for 30 s with gentle agitation.
3. Add 9 mL HBSS. Centrifuge for 5 min at 1,000 × *g*, 4°C.
4. Discard supernatant.

3.3. Dead Cell Depletion

The removal of dead cells from the cell suspension greatly aids in vivo tumour formation and downstream assays, such as Flow Cytometry analyses.

1. Centrifuge cell suspension for 5 min at 750 × *g*, 4°C and discard the supernatant.
2. For every 1×10^7 cells, add 100 μL of dead cell removal microbeads, mix well and incubate for 15 min at room temperature in dark (see Note 1).
3. Add 4 mL of 1× binding buffer and mix.
4. Filter suspension through a 40 μM filter (see Note 3).
5. Place the MACS® LD separation column in the MultiStand. Run 2 mL of 1× binding buffer through to prime the column. Discard the flowthrough.
6. Cells need to be at a density of $<50 \times 10^6$ cells/mL (see Note 4).
7. Apply up to 3 mL cells to the column and collect the flowthrough (see Note 5).
8. Apply 2 mL 1× binding buffer and collect flowthrough in sterile 15-mL tube. Repeat with another 2 mL of 1× binding buffer.
9. Centrifuge cell suspension for 5 min at 750 × *g*, 4°C and discard the supernatant. The pellet contains the live cell population.
10. Repeat dead cell labelling (steps 2–9) for a cleaner live fraction.
11. Resuspend pellet in PBS.
12. Count cell fraction using trypan blue staining.

3.4. CD31 and CD45 Depletion of Primary Endometrial Carcinomas

CD31 and CD45 are markers of endothelial and haematopoietic cells, respectively. They must be removed from a cell preparation before any subsequent studies on endometrial tumour-initiating cells can be undertaken. The majority of endometrial cancers are epithelial in origin, and the methods presented here pertain to those types only.

1. Centrifuge cells for 5 min at 1,000 × *g*, 4°C and discard the supernatant.
2. For every 1×10^7 cells, add 60 μL of HBSS/2% FBS/1 mM EDTA (see Note 6).
3. Add 20 μL of FcR blocking reagent, mix well, and incubate for 10 min at 4°C.
4. Add 10 μL of each microbead. Make up to a final volume of 100 μL with HBSS/2% FBS/1 mM EDTA. Incubate for 15 min in the dark at 4°C (see Note 7).
5. Add 2 mL of PBS/2% FBS/1 mM EDTA to cells, mix well, and centrifuge (260 × *g* for 10 min, 4°C).
6. Discard supernatant and repeat step 5.
7. Filter cells through a 40-μM filter (see Note 3).
8. Prepare the MACS® LD separation column by placing it in the MultiStand. Run 2 mL of PBS/2% FBS/1 mM EDTA through to prime the column. Discard the flowthrough.
9. The cells need to be at a density of $<50 \times 10^6$ cells/mL. Apply up to 3 mL cells to the column and collect the flowthrough.
10. Apply 2 mL PBS/2% FBS/1 mM EDTA and collect flowthrough. Repeat with another 2 mL of PBS/2% FBS/1 mM EDTA.
11. Centrifuge the collected flowthrough for 10 min at 260 × *g*, 4°C. This is the CD31/CD45 depleted cell population.
12. Resuspend cell pellet in 10 mL (or suitable volume) PBS/2% FBS/1 mM EDTA, trypan blue stain, and count.
13. If cells are to be injected into mice for in vivo tumourigenicity assays, then proceed to Subheading 3.5.

3.5. In Vivo Endometrial Cell Tumourigenicity

All experiments utilizing mouse models must be reviewed and approved by the host Institution's relevant animal care committee, and conform to appropriate national laws governing the care and use of laboratory animals. In this protocol, 6–12-week-old female non-obese diabetic/severe combined immunodeficient (NOD/SCID) mice are used for injections of human primary and xenograft-derived endometrial tumour epithelial cells. These mice are excellent hosts for xenograft development due to their deficiency in B- and T-cell lymphocyte development.

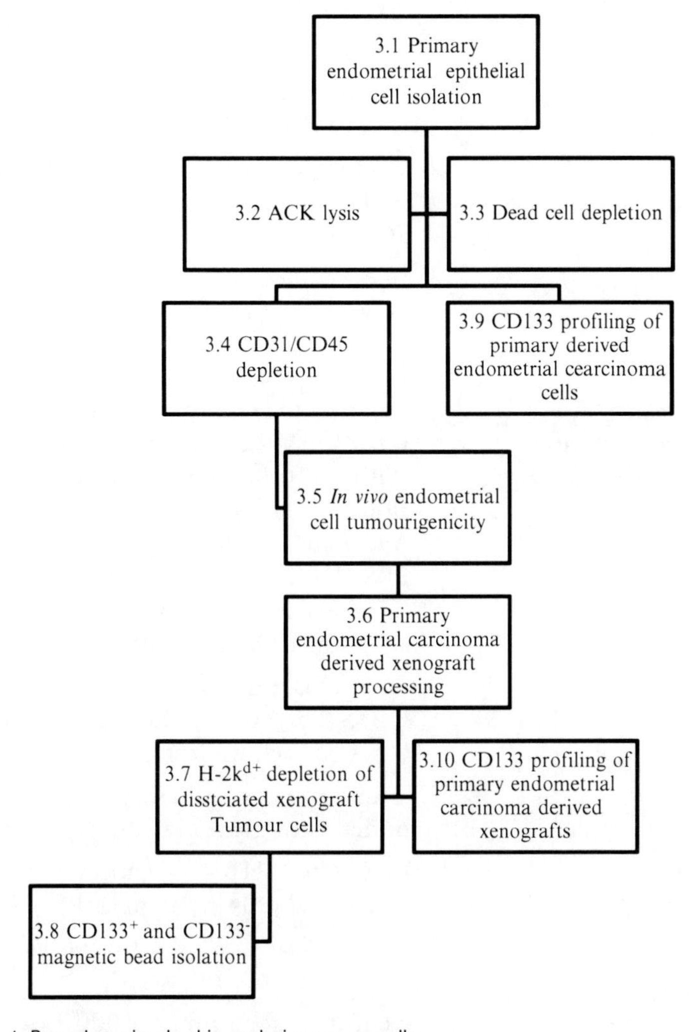

Fig. 1. Procedures involved in analyzing cancer cells.

If the potential tumourigenicity of a heterogenous population of tumour cells is to be investigated, cells can be injected in decreasing orders of magnitude (1×10^6 to 1×10^3). This approach is utilized for a variety of different cancers (9, 12). It has previously been reported that endometrial tumours contain a subpopulation of cells with tumour-initiating capabilities (3, 13, 14) that can be serially transplanted. Importantly, the transplanted xenografts and serially transplants xenografts maintain the same histological phenotype as that of the original primary tissue, confirming the value of such an in vivo model. These protocols describe the uses of this xenograft model and its application for the investigation of cancer cell properties (Fig. 1).

1. Suspend a defined number of isolated primary endometrial epithelial cells in 100 μL of sterile PBS (see Note 8). Add 100 μL of Matrigel® (see Note 9) and mix well. The ratio of PBS to Matrigel® is 1:1. Maintain on ice.
2. Using a 27-gauge needle subcutaneously inject this 200 μL solution of cells into the right and left dorsal sides of NOD/SCID mice.
3. Inject control animals simultaneously with 1:1 PBS/Matrigel® only. Assess tumour development daily.

3.6. Primary Endometrial Carcinoma-Derived Xenograft Processing

1. Isolate xenograft-derived tumours aseptically (see Note 10). Transfer to a 100-mm culture dish. Add 1 mL HBSS/2% FBS/1 mM EDTA. Use one blade to secure the tumour piece. With the other blade start at edge and chop finely, working along the entire tumour. Ideally mince to yield 2 mm^3 pieces. Add minced tumour to a 50-mL tube containing HBSS/2% FBS/1 mM EDTA, 1 mg/mL collagenase Type II/20 mL, 0.025% DNase I, and 1.5 mL amphotericin B. Incubate with agitation at 37°C for 30 min.
2. Filter cells through a 100-μm mesh filter and wash 3×5 min in HBSS/2% FBS/1 mM EDTA. Centrifuge at 1,000×*g* 4°C between each wash step.
3. Transfer cells to a fresh tube and incubate in 1 mL ACK lysis buffer, with gentle agitation, for 30 s at room temperature.
4. Add 9 mL HBSS. Centrifuge for 5 min at 1,000×*g* 4°C. Resuspend cells in 10 mL HBSS/2% FBS/1 mM EDTA.
5. Layer cells over 5 mL of Ficoll-Paque™ PLUS in a 15-mL falcon (see Note 11).
6. Centrifuge for 15 min, 1,000×*g* 4°C, with no brake and minimum acceleration.
7. Cell debris aggregates at the bottom of the tube. Collect the top supernatant and viable cells (visible as a grey/white layer), avoiding collection of any ficoll.
8. Wash 3×5 min in HBSS/2% FBS/1 mM EDTA. Centrifuge for 5 min, at 1,000×*g* 4°C between each wash step. Resuspend in 2 mL PBS/2% FBS/1 mM EDTA.
9. Trypan blue stain and count. Determine if a dead cell depletion step is needed.

3.7. H-2K^{d+} Depletion of Dissociated Xenograft Tumour Cells

The mouse major compatibility complex (MHC) antigen (H-2) includes the classical MHC class 1 subclass H-2K^{d} found in NOD/SCID mice. Thus, murine cells can be eliminated from a sample preparation by H-2K^{d+} magnetic bead depletion.

1. Centrifuge cells for 5 min at 1,000 × *g* 4°C. Discard the supernatant (see Note 6).
2. For every 1×10^7 cells, add 60 μL of PBS/2% FBS/1 mM EDTA to the cell pellet. Add 20 μL of FcR blocking reagent, mix well, and incubate for 10 min at 4°C in dark.
3. Add 20 μL of H-2K^d FITC-conjugated antibody. Make up to 100 μL with PBS/2% FBS/1 mM EDTA. Mix and incubate for 10 min at 4°C in dark. If incubation is on ice, increase the time to 30 min.
4. Add 1 mL of PBS/2% FBS/1 mM EDTA, mix well, and centrifuge (260 × *g* for 10 min, 4°C).
5. Resuspend pellet in 90 μL of PBS/2% FBS/1 mM EDTA. Add 10 μL of anti-FITC microbeads. Incubate for 15 min at 4°C in dark. Wash with 1 mL of PBS/2% FBS/1 mM EDTA and centrifuge (260 × *g* for 10 min, 4°C). Resuspend in 2 mL of PBS/2% FBS/1 mM EDTA.
6. Filter through a 40 μM filter to remove clumps that may clog the column. The cells need to be at a density of $<50 \times 10^6$ cells/mL; otherwise, they will block the LD column. However, the minimal volume to be used is 2 mL.
7. Prepare the LD column by placing it in the MultiStand. Run 2 mL of PBS/2% FBS/1 mM EDTA through to prime the column. Discard the flowthrough.
8. Apply up to 3 mL cells to the column and collect the flowthrough.
9. Apply 2 mL of PBS/2% FBS/1 mM EDTA and collect the flowthrough. Repeat with another 2 mL of PBS/2% FBS/1 mM EDTA.
10. Centrifuge the collected flowthrough (260 × *g*, 10 min at 4°C). This is the depleted (H-2K^{d-}) cell population. Resuspend cell pellet in 10 mL of PBS/2% FBS/1 mM EDTA, trypan blue stain, and count (see Note 12).

3.8. CD133$^+$ and CD133$^-$ Magnetic Bead Isolation

CD133 (human Prominin-1, AC133), originally identified in haematopoietic stem cells (18, 19), is also expressed on primitive cells of neural, endothelial, and epithelial lineages. Several investigators have identified CD133 as a potential tumour-initiating cell marker in solid tumours of the brain (8), prostate (20), colon (9) ovary (11, 12), and more recently the endometrium (13). CD133$^+$ cells have been associated with an increase in in vivo tumour initiation (8, 11, 13), asymmetric cell division, and increased resistance to chemotherapeutic drugs (11), as compared to CD133$^-$ cells. The following is a method used for isolating both CD133$^+$ and CD133$^-$ cells from H-2K^{d-} cells by magnetic bead selection.

1. Centrifuge cells at $650 \times g$ and discard the supernatant.
2. For every 1×10^7 cells, add 20 μL of FcR blocking reagent and 60 μL PBS/2% FBS/1 mM EDTA, mix well, and incubate for 10 min at 4°C in dark.
3. For every 1×10^7 cells, add 10 μL of PBS/2% FBS/1 mM EDTA to the cell pellet. Add 100 μL of CD133 microbeads, mix well, and incubate for 15 min at 4°C in dark.
4. Add 2 mL of PBS/2% FBS/1 mM EDTA and centrifuge ($260 \times g$ for 10 min, 4°C).
5. Discard supernatant. Repeat step 4.
6. Filter cell suspension through a 40-μM filter.
7. Prepare the MACS® LD separation column by placing it in the MultiStand. Run 2 mL of PBS/2% FBS/1 mM EDTA through to prime the column. Discard the flowthrough.
8. The cells need to be at a density of $<50 \times 10^6$ cells/mL. Apply up to 3 mL cells in to the column and collect the flowthrough.
9. Apply 2 mL PBS/2% FBS/1 mM EDTA and collect flowthrough. Repeat with another 2 mL of PBS/2% FBS/1 mM EDTA.
10. Centrifuge ($600 \times g$ for 10 min, 4°C). Collect the flowthrough. This is the $CD133^-$ population.
11. Apply the $CD133^-$ population to a new primed column. Collect as before, this is the $CD133^-$ pure population (see Note 13).
12. To recover the $CD133^+$ population, remove column from magnet. Place over a 15-mL tube. Add 2 mL PBS/2% FBS/1 mM EDTA, plunge (plunger is supplied with the column). Repeat with additional 2 mL PBS/2% FBS/1 mM EDTA.
13. Trypan blue stain and count cell fractions. Inject cell populations (see Subheading 3.5) and monitor tumour growth daily.

3.9. CD133 Profiling of Primary-Derived Endometrial Carcinoma Cells

This protocol uses epithelial cells isolated from primary endometrial carcinomas. Unless isolated cells are to be used for in vivo injections, CD31 and CD45 bead depletion is not necessary, as such positive cells will be excluded from analyses based on the following staining methods:

1. Ensure cells are in a single suspension. This protocol requires 8×1.5-mL Eppendorf tubes. Maintain cells on ice unless otherwise stated.
2. For every 1×10^7 cells, add 80 μL PBS/2% FBS/1 mM EDTA and 20 μL of FcR Blocking Reagent, mix well, and incubate for 10 min at 4°C in dark.
3. Bring to a suitable volume with PBS/2% FBS/1 mM EDTA (see Note 14).

Table 1
Staining procedure for primary endometrial carcinoma cells

Tube number	1	2	3	4	5	6	7	8
Buffer	110 µL	[a]	[a]	[a]	[a]	[a]	[a]	[a]
CD44-APC	–	–	–	–	–	–	5 µL	–
CD31-FITC	–	10 µL	–	10 µL	10 µL	–	–	–
CD45-FITC	–	20 µL	–	20 µL	20 µL	5 µL	–	–
IgG2b-PE	–	–	–	3.57 µL	–	–	–	–
CD133/2-PE	–	–	–	10 µL	10 µL	–	–	5 µL

[a]See Note 14

4. Add 0.5×10^6 cells to each of tubes 1–4. Add 2×10^6 cells to tube 5 (see Note 15).
5. Stain cells immediately in accordance with Table 1 (see Note 16).
6. Incubate cells for 10 min at 4°C in dark.
7. After staining, add 1 mL PBS/2% FBS/1 mM EDTA to cells and centrifuge ($600 \times g$ for 10 min, 4°C). Discard flowthrough. Resuspend cells in 1 mL PBS.
8. Add 1 µL LIVE/DEAD working solution to tubes 3–5 (see Note 17). Incubate for 30 min at 4°C in dark.
9. Meanwhile, add BD™ CompBeads (one drop of negative control and one drop of anti-mouse Ig, κ beads to 100 µL of PBS/2% FBS/1 mM EDTA) (see Note 18) to tubes 6–8 and stain with conjugated antibodies, as outlined in Table 1. Incubate for 25 min at room temperature in dark. Add 200 µL of PBS/2% FBS/1 mM EDTA to tubes 6–8. Do not add PFA. Store at 4°C, in dark, until ready for analysis.
10. Centrifuge tubes 3–5 ($260 \times g$ for 10 min, 4°C). Discard flowthrough. Resuspend cells in 200 µL of 4% PFA. Incubate for 1 h at 4°C in dark. Then, centrifuge ($600 \times g$ for 10 min, 4°C) and resuspend cells in 500 µL PBS/2% FBS/1 mM EDTA. Store at 4°C, in dark, until ready for analysis. In this method, a LSRII (BD Biosciences) was used for surface marker analyses.

3.10. CD133 Profiling of Primary Endometrial Carcinoma-Derived Xenografts

This protocol uses cells isolated from primary endometrial carcinomas-derived xenografts.

1. Ensure cells are in a single suspension. This protocol requires 7×1.5-mL Eppendorf tubes. Maintain cells on ice unless otherwise stated.

Table 2
Staining procedure for primary endometrial carcinoma-derived xenograft cells

Tube number	1	2	3	4	5	6	7
Buffer	110 μL	[a]	[a]	[a]	[a]	[a]	[a]
H-2K^d-FITC	–	2 μL	2 μL	2 μL	2 μL	–	–
CD44-APC	–	–	–	–	–	5 μL	–
IgG2b-PE	–	–	3.57 μL	–	–	–	–
CD133/2-PE	–	–	–	10 μL	–	–	5 μL

[a]See Note 14

2. For every 1×10^7 cells, add 80 μL PBS/2% FBS/1 mM EDTA and 20 μL of FcR Blocking Reagent, mix well, and incubate for 10 min at 4°C in dark.
3. Bring to a suitable volume with PBS/2% FBS/1 mM EDTA (see Note 14).
4. Add 0.5×10^6 cells to each of tubes 1–3. Add 2×10^6 cells to tube 4 (see Note 15).
5. Stain cells immediately in accordance with Table 2 (see Note 19).
6. Incubate cells for 10 min at 4°C in dark.
7. After staining, add 1 mL PBS/2% FBS/1 mM EDTA to cells and centrifuge ($600 \times g$ for 10 min, 4°C). Discard flowthrough. Resuspend cells in 1 mL PBS.
8. Add 1 μL LIVE/DEAD working solution to tubes 3–4. Incubate for 30 min at 4°C in dark (see Note 18).
9. Meanwhile, add BD™ CompBeads (one drop of Negative Control and one drop of anti-mouse Ig, κ beads to 100 μL of PBS/2% FBS/1 mM EDTA) (see Note 15) to tubes 5–7 and stain with conjugated antibodies, as outlined in Table 2. Incubate for 25 min at room temperature in dark. Add 200 μL of PBS/2% FBS/1 mM EDTA to tubes 5–7. Do not add PFA. Store at 4°C, in dark, until ready for analysis.
10. Centrifuge tubes 1–4 ($260 \times g$ for 10 min, 4°C). Discard flowthrough. Resuspend cells in 200 μL of 4% PFA. Incubate for 1 h at 4°C in dark. Then, centrifuge ($260 \times g$ for 10 min, 4°C) resuspend cells in 500 μL PBS/2% FBS/1 mM EDTA and store at 4°C in dark until ready for analysis. An LSRII (BD Biosciences) was used for surface marker analyses for these samples.

3.11. Murine Bone Marrow Isolation: Disseminated Tumour Cell Detection

Evidence suggests that there exist bone marrow niches that regulate the survival, proliferation, and differentiation of haematopoietic stem cells. At least two distinct haematopoietic stem cell-supportive niches have been identified in the bone marrow (21–23): an osteoblastic (endosteal) niche (24–26) and a perivascular niche (27, 28). Tumour cells [sometimes referred to as disseminated tumour cells, (DTCs)] derived from xenografts have been detected in both such niches (29).

1. Sacrifice mouse by CO_2 inhalation (see Note 20). Place mouse on its front. Spray hind legs with 70% ethanol. Turn mouse over. Spray abdomen and hind legs with 70% ethanol.
2. Using sterile scissors, carefully remove all skin from the hind legs, taking care not to cut into the abdomen or peritoneum of the mouse. Cut both legs off above the hip joint. Cut each foot off above the ankle joint.
3. Using a separate sterile scissors, remove all of the muscle surrounding the hind legs. Keep all remaining steps on ice where possible.
4. Place leg bones (femur, tibia, and fibula) in individual wells of a 6-well culture dish, containing 2–3 mL of sterile PBS. Keep legs from individual mice separate.
5. Place individual mice legs (i.e. both hind legs) in a sterile mortar. Transfer the PBS (2–3 mL/well) from the 6-well plate into the mortar. Crush with pestle (see Note 21).
6. Transfer crushed bone and bone marrow to a 50-mL tube via a 40-μM filter, using a 25-mL pipette. Ensure that all bone and bone marrow is removed from the mortar. Use PBS as a wash. The final volume of PBS is 30-mL/50-mL tube.
7. Centrifuge for 7 min at $200 \times g$, 4°C.
8. Resuspend pellet in 10 mL ACK lysis buffer. Incubate at room temperature for 5 min.
9. Add 5 mL PBS. Centrifuge for 7 min at $200 \times g$, 4°C. Discard the supernatant.
10. Resuspend cell pellet in 1 mL PBS.
11. Trypan blue stain and count (see Note 22). Cells may also be used for RNA isolation using TriReagent or for DNA extraction (see Subheading 3.12.4; resuspend 5×10^6 cells in 200 μL of PBS, add 20 μL proteinase K (supplied in Qiagen DNeasy Blood & Tissue Kit and proceed from step 3)).

3.12. CTC Detection

CTCs are cells detected in the peripherial blood that originated from either the primary tumour or from its metastases. DTCs (tumour cells detected in bone marrow; see Subheading 3.11) are

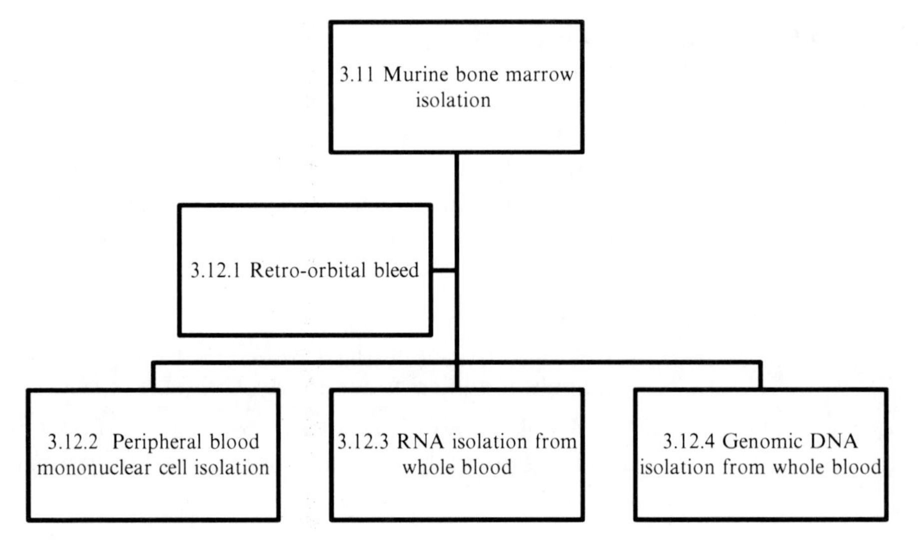

Fig. 2. Procedures involved in analyzing circulating tumour cells.

thought to reflect the metastatic potential of primary tumours. The presence of both CTCs (30, 31) and DTCs (32, 33) has been correlated to cancer patient prognosis and survival. In addition, peripheral blood mononuclear cells are sources of primitive haematopoietic progenitor cells and circulating CD133$^+$ endothelial progenitor cells are predictive of recurrence in colon cancer (34) and bone metastases (35). The following are methods for isolating cells from peripheral blood. Included are methods for nucleic acid isolation from whole blood (Fig. 2).

3.12.1. Retro-orbital Bleed

In the USA, the most common rodent-bleeding method is retro-orbital, puncturing the orbital sinus behind the eye. This technique will yield 0.5–1 mL blood, but its use is strictly managed, so ensure that the correct legislation is followed as there are other methods available for drawing rodent blood. Mice must be anaesthetized before attempting an orbital bleed.

1. Using a 27-gauge needle, inject Avertin (200 mg/kg body weight) intraperitoneal (IP). Take care to monitor the mouse during this time period. Perform a "toe-pinch" to ensure that the mouse is fully anaesthetized.
2. Hold the mouse against a flat surface. Use your forefinger to pull the facial skin taut while grasping the skin at the back of the neck. This will make the eye protrude slightly.
3. Gently insert the tip of the heparinized capillary into the corner of the eye at an angle of approximately 45°. You will feel the capillary meet resistance. Gently twist the capillary between

your thumb and forefinger. Once the sinus is ruptured blood flows into the capillary. As the mouse is generally sacrificed following an orbital bleed, both eyes can be bled.

4. Transfer blood to an EDTA-coated collection tube (for nucleic acid isolation or mononuclear cell isolation) and then to a 2-mL tube containing 1.3 mL of RNA *later*™ (for future RNA isolation) (see Note 23).
5. Sacrifice the mouse by the protocol approved by your Institution.

3.12.2. Peripheral Blood Mononuclear Cell Isolation

1. Layer approximately 700 μL of isolated whole blood on top of 5 mL Ficoll-Paque™ PLUS in a 15-mL tube. Rinse the EDTA-coated tube with sterile PBS to remove any remaining blood (see Note 24).
2. Layer PBS/blood washing into Ficoll-Paque™ PLUS tube in step 1. Ensure that final volume in tube is 7 mL (5 mL of Ficoll-Paque™ PLUS and 2 mL of blood/PBS).
3. Leave at room temperature for 5 min.
4. Centrifuge for 15 min at 1,000 × *g*, room temperature, with no brake and minimum acceleration.
5. Using a sterile Pasteur pipette, carefully remove the mononuclear cell layer. This layer appears as a thin white-coloured layer between the bottom layer (ficoll, pelleted red blood cells and granulocytes) and the top layer (plasma and platelets). The plasma may be stored for further analysis.
6. Transfer the mononuclear layer to another sterile 15-mL tube. Add 5 mL sterile PBS.
7. Centrifuge for 15 min at 1,000 × *g*, room temperature (see Note 25).
8. If red blood cell contamination is suspected, the cells may be treated with ACK lysis buffer here.
9. Resuspend pellet in 500 μL of PBS. Trypan blue stain and count. If cells are to be injected into mice do so after this step. On average, 3×10^6 cells will be recovered by this method. Cells should be injected in 300 μL of PBS via the tail vein (see Note 22 for injection method). Alternatively, total RNA may be isolated.

3.12.3. RNA Isolation from Whole Blood

This procedure isolates total RNA from whole blood collected by orbital bleeding. All reagents are supplied in the recommended kit (RiboPure Blood Kit).

1. Centrifuge blood sample collected in RNA later for 1 min, at 12,000 × *g*, room temperature. Discard the supernatant.

2. Do all the remaining steps in a chemical hood as the reagents used are toxic.
3. Add 800 μL lysis solution and 50 μL sodium acetate. Vortex. Add 500 μL acid-phenol: chloroform. Vortex.
4. Incubate for 5 min, at room temperature.
5. Centrifuge for 1 min, at 12,000 × *g*, room temperature. Transfer the top aqueous layer to a new 1.5-mL tube. Add 600 μL 100% ethanol. Vortex.
6. Apply sample to filter cartridge assembly in 700-μL aliquots.
7. Centrifuge for 10 s, at 12,000 × *g*, room temperature. Discard flowthrough.
8. Repeat steps 6 and 7 until all sample is passed through column.
9. Apply 700 μL wash solution 2/3 to filter column. Centrifuge for 10 s, at 11.7 × *g*, room temperature. Discard flowthrough.
10. Repeat step 9.
11. Transfer filter column to new collection tube.
12. Apply 50 μL elution solution (pre-heated to 75°C). Incubate for 5 min at room temperature.
13. Centrifuge for 30 s, at 12,000 × *g*, room temperature.
14. Repeat step 12. Centrifuge for 1 min, at 12,000 × *g*, room temperature.
15. Store eluted RNA at –80°C.

3.12.4. Genomic DNA Isolation from Whole Blood

All reagents and buffers with the exception of 100% ethanol are provided in the recommended kit (Qiagen DNeasy Blood & Tissue Kit).

1. Add 20 μL proteinase K (600 mAU/mL) to a 1.5-mL tube.
2. Add 100 μL anti-coagulated blood and 100 μL PBS. Final volume is now 220 μL.
3. Add 200 μL buffer AL. Mix and incubate for 10 min at 56°C.
4. Add 200 μL 100% ethanol. Mix. Apply sample to DNeasy mini column with a 2-mL collection tube.
5. Centrifuge for 1 min at 2,000 × *g*, room temperature. Discard flowthrough and collection tube.
6. Place column into a new 2-mL collection tube. Add 500 μL buffer AW1. Centrifuge for 3 min at 12,000 × *g*, room temperature. Discard flowthrough and collection tube.
7. Place column into a 1.5-mL tube. Add 200 μL buffer AE. Incubate for 1 min at room temperature.
8. Centrifuge for 1 min at 2,000 × *g*, room temperature.
9. Store eluted DNA at –80°C.

4. Notes

1. Unless otherwise stated, all procedures must be undertaken aseptically if possible.
2. Stroma cells are in the supernatant, and epithelial cells are in the pellet. If stroma cells are required (e.g. if a sarcoma is being processed), after the 10 min incubation has passed, split the resulting supernatant between two 15-mL tubes. Centrifuge both tubes for 5 min at $750 \times g$, 4°C. Decant the supernatant. The pellet contains the endometrial stroma cells.
3. This step is crucial to prevent blockage of the column.
4. This prevents blockage of column.
5. Due to the column structure, the minimal volume to be used is 2 mL. Different separation column types have different volume capacities.
6. All procedures must be undertaken aseptically. This protocol is for up to 1×10^7 cells. If cell number is lower, use same reagent volumes. If there are greater than 1×10^7 cells, adjust reagent volumes and total volumes accordingly.
7. If incubation is on ice, increase the time by 20 min. The CD31 and CD45 microbeads are light sensitive and must be kept in the dark.
8. The cell number injected depends on the application of the in vivo study. For generating tumour cells for additional characterization studies, it is desirable to inject sufficient cell numbers. Also bear in mind that as the xenografts progress through the transplant stages, it will take fewer cells to generate tumours in vivo. However, care must be taken not to allow the transplant number to go beyond 3 or 4 as this positively selects for "super" tumour-initiating cells.
9. Matrigel® solidifies at room temperature. Thaw matrigel® on ice before use and maintain on ice for the duration of the injection preparation. Additionally, chill the 27-gauge needles/syringes on ice prior to injections.
10. Tumours should be harvested when they are ≤10 mm in diameter. If they are allowed to grow larger, then the inside will become necrotic. It is recommended to keep a portion of the xenograft tissue for 4% PFA fixing, OCT mounting, dry frozen, and in 10% DMSO/medium. This will enable future immunohistochemical analyses on FFPE and frozen sections, nucleic acid/protein analyses, and cell line derivation if possible.
11. If $>50 \times 10^6$ cells, use 2×15-mL tubes (10 mL of cell suspension is layered on top of each 5 mL of Ficoll-Paque™).

12. Freeze back a portion of H-2K^{d-} cells in liquid nitrogen if possible. This will enable future nucleic acid/protein analyses.
13. The addition of this step enables a very pure CD133^{-} fraction to be isolated.
14. The final volume per tube must not exceed 110 μL. Excluding tube 1, the volume of PBS/2% FBS/1 mM EDTA added is dependent on cell concentration and the antibodies used.
15. This number can be increased or reduced depending on how many events will be captured by Flow analyses.
16. The appropriate concentration of conjugated antibody used in staining cells needs to be determined initially by analyses of varying concentrations of antibody. This is achieved by staining with serial dilutions of antibodies and their corresponding isotype controls. For this assay, using this tumour type, we determined the final concentration of antibody to be 0.15 μg/110 μL for CD133/2-PE. Hence, the corresponding IgG2b-PE isotype control was used at a final concentration of 0.15 μg/110 μL.
17. The final concentration of LIVE/DEAD to use needs to be determined by a serial titration. It differs for each tumour type.
18. One drop equates to approximately 60 μL.
19. The appropriate concentration of conjugated antibody used in staining cells needs to be determined initially by analyses of varying concentrations of antibody. This is achieved by staining with serial dilutions of antibodies and their corresponding isotype controls. For this assay, using this tumour type, we determined the final concentration of antibodies to be 1 μg/110 μL (H-2K^{d}-FITC) and 0.15 μg/110 μL (CD133/2-PE). Hence, the corresponding IgG2b-PE isotype control was used at a final concentration of 0.15 μg/110 μL.
20. Follow the approved method of animal sacrifice as outlined by the Institution.
21. To ensure that tumour circulating cells present in either of the bone marrow niches are isolated, the bones are crushed along with the marrow, instead of the marrow being flushed from the bone. After crushing, if a thin red line is visible in the centre of the bone, it indicates that not all of the marrow has been removed. Maintain crushing until a pink homogenous solution remains.
22. If isolated bone marrow cells are to be injected into mice via the tail vein, do so after this step. For example, if labelled cells (e.g. GFP tagged) or cells with a known mutation are to be tracked in a recipient inject 25×10^6 cells (using an insulin needle bent to a 90° angle) in 400 μL of sterile PBS via the tail vein. Ensure that the mouse has been placed under a heating

lamp for ~10 min to ensure veins are completely dilated. Ear tag the mouse and keep a record of host mouse to recipient mouse identity.

23. Blood can be transferred to as many EDTA tubes as necessary. Five hundred microlitres of blood in 1.3 mL RNA *later*™ works well.
24. Addition of PBS also dilutes the blood which prevents lymphocyte–erythrocytes aggregates from forming.
25. This wash step removes any remaining platelets that may be in the mononuclear layer.

Acknowledgements

Preparation of this chapter was financed by the Science Foundation Ireland Strategic Research Cluster, Molecular Therapeutics for Cancer, Ireland (08/SRC/B1410: L. O'D. and J.C.). The authors thank Dr. Bo Rueda and staff at the Vincent Center for Reproductive Biology, Massachusetts General Hospital, Boston, USA, where this work was undertaken.

References

1. Dalerba, P., Cho, R. W., and Clarke, M. F. (2007) Cancer stem cells: models and concepts, *Annu Rev Med* **58**, 267–284.
2. Kelly, P. N., Dakic, A., Adams, J. M., Nutt, S. L., and Strasser, A. (2007) Tumor growth need not be driven by rare cancer stem cells, *Science* **317**, 337.
3. Friel, A. M., Sergent, P. A., Patnaude, C., Szotek, P. P., Oliva, E., Scadden, D. T., Seiden, M. V., Foster, R., and Rueda, B. R. (2008) Functional analyses of the cancer stem cell-like properties of human endometrial tumor initiating cells, *Cell Cycle* 7, 242–249.
4. Haraguchi, N., Utsunomiya, T., Inoue, H., Tanaka, F., Mimori, K., Barnard, G. F., and Mori, M. (2006) Characterization of a side population of cancer cells from human gastrointestinal system, *Stem Cells* **24**, 506–513.
5. Patrawala, L., Calhoun, T., Schneider-Broussard, R., Zhou, J., Claypool, K., and Tang, D. G. (2005) Side population is enriched in tumorigenic, stem-like cancer cells, whereas ABCG2+ and ABCG2– cancer cells are similarly tumorigenic, *Cancer Res* **65**, 6207–6219.
6. Szotek, P. P., Pieretti-Vanmarcke, R., Masiakos, P. T., Dinulescu, D. M., Connolly, D., Foster, R., Dombkowski, D., Preffer, F., Maclaughlin, D. T., and Donahoe, P. K. (2006) Ovarian cancer side population defines cells with stem cell-like characteristics and Mullerian Inhibiting Substance responsiveness, *Proc Natl Acad Sci U S A* **103**, 11154–11159.
7. Al-Hajj, M., Wicha, M. S., Benito-Hernandez, A., Morrison, S. J., and Clarke, M. F. (2003) Prospective identification of tumorigenic breast cancer cells, *Proc Natl Acad Sci U S A* **100**, 3983–3988.
8. Singh, S. K., Hawkins, C., Clarke, I. D., Squire, J. A., Bayani, J., Hide, T., Henkelman, R. M., Cusimano, M. D., and Dirks, P. B. (2004) Identification of human brain tumour initiating cells, *Nature* **432**, 396–401.
9. Ricci-Vitiani, L., Lombardi, D. G., Pilozzi, E., Biffoni, M., Todaro, M., Peschle, C., and De Maria, R. (2007) Identification and expansion of human colon-cancer-initiating cells, *Nature* **445**, 111–115.
10. Zhang, S., Balch, C., Chan, M. W., Lai, H. C., Matei, D., Schilder, J. M., Yan, P. S., Huang, T. H., and Nephew, K. P. (2008) Identification and characterization of ovarian cancer-initiating cells from primary human tumors, *Cancer Res* **68**, 4311–4320.

11. Baba, T., Convery, P. A., Matsumura, N., Whitaker, R. S., Kondoh, E., Perry, T., Huang, Z., Bentley, R. C., Mori, S., Fujii, S., Marks, J. R., Berchuck, A., and Murphy, S. K. (2009) Epigenetic regulation of CD133 and tumorigenicity of CD133+ ovarian cancer cells, *Oncogene* **28**, 209–218.
12. Curley, M. D., Therrien, V. A., Cummings, C. L., Sergent, P. A., Koulouris, C. R., Friel, A. M., Roberts, D. J., Seiden, M. V., Scadden, D. T., Rueda, B. R., and Foster, R. (2009) CD133 Expression Defines a Tumor Initiating Cell Population in Primary Human Ovarian Cancer, *Stem Cells* 27(12), 2875–2883.
13. Rutella, S., Bonanno, G., Procoli, A., Mariotti, A., Corallo, M., Prisco, M. G., Eramo, A., Napoletano, C., Gallo, D., Perillo, A., Nuti, M., Pierelli, L., Testa, U., Scambia, G., and Ferrandina, G. (2009) Cells with characteristics of cancer stem/progenitor cells express the CD133 antigen in human endometrial tumors, *Clin Cancer Res* **15**, 4299–4311.
14. Hubbard, S. A., Friel, A. M., Kumar, B., Zhang, L., Rueda, B. R., and Gargett, C. E. (2009) Evidence for cancer stem cells in human endometrial carcinoma, *Cancer Res* **69**, 8241–8248.
15. Bansal, N., Yendluri, V., and Wenham, R. M. (2009) The molecular biology of endometrial cancers and the implications for pathogenesis, classification, and targeted therapies, *Cancer Control* **16**, 8–13.
16. Eliane, J. P., Repollet, M., Luker, K. E., Brown, M., Rae, J. M., Dontu, G., Schott, A. F., Wicha, M., Doyle, G. V., Hayes, D. F., and Luker, G. D. (2008) Monitoring serial changes in circulating human breast cancer cells in murine xenograft models, *Cancer Res* **68**, 5529–5532.
17. Rago, C., Huso, D. L., Diehl, F., Karim, B., Liu, G., Papadopoulos, N., Samuels, Y., Velculescu, V. E., Vogelstein, B., Kinzler, K. W., and Diaz, L. A., Jr. (2007) Serial assessment of human tumor burdens in mice by the analysis of circulating DNA, *Cancer Res* **67**, 9364–9370.
18. Yin, A. H., Miraglia, S., Zanjani, E. D., Almeida-Porada, G., Ogawa, M., Leary, A. G., Olweus, J., Kearney, J., and Buck, D. W. (1997) AC133, a novel marker for human hematopoietic stem and progenitor cells, *Blood* **90**, 5002–5012.
19. Miraglia, S., Godfrey, W., Yin, A. H., Atkins, K., Warnke, R., Holden, J. T., Bray, R. A., Waller, E. K., and Buck, D. W. (1997) A novel five-transmembrane hematopoietic stem cell antigen: isolation, characterization, and molecular cloning, *Blood* **90**, 5013–5021.
20. Collins, A. T., Berry, P. A., Hyde, C., Stower, M. J., and Maitland, N. J. (2005) Prospective identification of tumorigenic prostate cancer stem cells, *Cancer Res* **65**, 10946–10951.
21. Lane, S. W., Scadden, D. T., and Gilliland, D. G. (2009) The leukemic stem cell niche: current concepts and therapeutic opportunities, *Blood* **114**, 1150–1157.
22. Iwasaki, H., and Suda, T. (2009) Cancer stem cells and their niche, *Cancer Sci* **100**, 1166–1172.
23. Meads, M. B., Hazlehurst, L. A., and Dalton, W. S. (2008) The bone marrow microenvironment as a tumor sanctuary and contributor to drug resistance, *Clin Cancer Res* **14**, 2519–2526.
24. Adams, G. B., Chabner, K. T., Alley, I. R., Olson, D. P., Szczepiorkowski, Z. M., Poznansky, M. C., Kos, C. H., Pollak, M. R., Brown, E. M., and Scadden, D. T. (2006) Stem cell engraftment at the endosteal niche is specified by the calcium-sensing receptor, *Nature* **439**, 599–603.
25. Mayack, S. R., and Wagers, A. J. (2008) Osteolineage niche cells initiate hematopoietic stem cell mobilization, *Blood* **112**, 519–531.
26. Zhang, J., Niu, C., Ye, L., Huang, H., He, X., Tong, W. G., Ross, J., Haug, J., Johnson, T., Feng, J. Q., Harris, S., Wiedemann, L. M., Mishina, Y., and Li, L. (2003) Identification of the haematopoietic stem cell niche and control of the niche size, *Nature* **425**, 836–841.
27. Kiel, M. J., Yilmaz, O. H., Iwashita, T., Terhorst, C., and Morrison, S. J. (2005) SLAM family receptors distinguish hematopoietic stem and progenitor cells and reveal endothelial niches for stem cells, *Cell* **121**, 1109–1121.
28. Kiel, M. J., and Morrison, S. J. (2008) Uncertainty in the niches that maintain haematopoietic stem cells, *Nat Rev Immunol* 8, 290–301.
29. Ninomiya, M., Abe, A., Katsumi, A., Xu, J., Ito, M., Arai, F., Suda, T., Kiyoi, H., Kinoshita, T., and Naoe, T. (2007) Homing, proliferation and survival sites of human leukemia cells in vivo in immunodeficient mice, *Leukemia* **21**, 136–142.
30. Cristofanilli, M., Hayes, D. F., Budd, G. T., Ellis, M. J., Stopeck, A., Reuben, J. M., Doyle, G. V., Matera, J., Allard, W. J., Miller, M. C., Fritsche, H. A., Hortobagyi, G. N., and Terstappen, L. W. (2005) Circulating tumor cells: a novel prognostic factor for newly diagnosed metastatic breast cancer, *J Clin Oncol* **23**, 1420–1430.
31. Smerage, J. B., and Hayes, D. F. (2008) The prognostic implications of circulating tumor cells in patients with breast cancer, *Cancer Invest* **26**, 109–114.

32. Braun, S., Pantel, K., Muller, P., Janni, W., Hepp, F., Kentenich, C. R., Gastroph, S., Wischnik, A., Dimpfl, T., Kindermann, G., Riethmuller, G., and Schlimok, G. (2000) Cytokeratin-positive cells in the bone marrow and survival of patients with stage I, II, or III breast cancer, *N Engl J Med* **342**, 525–533.
33. Gebauer, G., Fehm, T., Merkle, E., Beck, E. P., Lang, N., and Jager, W. (2001) Epithelial cells in bone marrow of breast cancer patients at time of primary surgery: clinical outcome during long-term follow-up, *J Clin Oncol* **19**, 3669–3674.
34. Lin, E. H., Hassan, M., Li, Y., Zhao, H., Nooka, A., Sorenson, E., Xie, K., Champlin, R., Wu, X., and Li, D. (2007) Elevated circulating endothelial progenitor marker CD133 messenger RNA levels predict colon cancer recurrence, *Cancer* **110**, 534–542.
35. Mehra, N., Penning, M., Maas, J., Beerepoot, L. V., van Daal, N., van Gils, C. H., Giles, R. H., and Voest, E. E. (2006) Progenitor marker CD133 mRNA is elevated in peripheral blood of cancer patients with bone metastases, *Clin Cancer Res* **12**, 4859–4866.

Chapter 6

Gene Expression Profiling in Formalin-Fixed, Paraffin-Embedded Tissues Using the Whole-Genome DASL Assay

Craig S. April and Jian-Bing Fan

Abstract

Here, we provide a detailed technical description of a gene expression assay (Whole-Genome DASL (WG-DASL)), which not only enables whole-genome transcriptional profiling of degraded material, such as formalin-fixed, paraffin-embedded tissues, but is also capable of generating robust profiles with low input intact RNA. The WG-DASL assay combines target-specific annealing, extension, and ligation events followed by universal PCR and labeling steps to generate highly multiplexed Cy3-labeled products. These short products, which are single-stranded, are directly hybridized to a whole-genome expression BeadChip (HumanRef-8) containing probe content corresponding to ~24 K RefSeq transcripts. After washing and imaging, fluorescence emissions are quantitatively recorded for each probe using high-resolution confocal scanners and imaging software. GenomeStudio software allows direct analysis of mRNA expression data and provides results in standard file formats that can be readily exported and analyzed with most standard gene expression analysis software programs. This technology is particularly useful for genome-wide expression profiling in degraded, archived material, including limited quantities of clinical samples, such as microdissected and biopsied materials.

Key words: Formalin-fixed, paraffin-embedded tissues, Archived samples, RNA, Gene expression analysis, Microarray, BeadArray, DASL assay, Biomarker

1. Introduction

1.1. BeadArray Technology

The manufacture of Illumina microarrays has been described elsewhere (1). Briefly, the WG-DASL assay uses a multisample BeadChip (HumanRef-8) as a readout, which is composed of eight individual arrays manufactured on a single microscope slide-shaped silicon substrate. The content of the multisample HumanRef-8 BeadChip is derived from the NCBI RefSeq database (Build 36.2, Release 22), which includes >24,000 hand-curated, well-annotated

Lorraine O'Driscoll (ed.), *Gene Expression Profiling: Methods and Protocols*, Methods in Molecular Biology, vol. 784,
DOI 10.1007/978-1-61779-289-2_6, © Springer Science+Business Media, LLC 2011

transcripts, corresponding to ~18 K unique genes. BeadChips are constructed by introducing oligonucleotide-bearing 3 μm beads into microwells etched into the surface of the silicon substrate. During the manufacturing process, beads self-assemble into the microwells of the BeadChips. Each bead contains hundreds of thousands of copies of covalently attached, full-length oligonucleotide probes, and is represented with an average 30-fold redundancy on every BeadChip. After random bead assembly, 29-mer address sequences present on each bead are used for a hybridization-based procedure to map the array, identifying the location of each bead type. This final process validates the hybridization performance of every bead on every BeadChip, providing 100% array QC. IntelliHyb® seal technology allows different samples to be hybridized to a single BeadChip, simplifying the processing and streamlining the workflow. All steps downstream of the hybridization are performed in parallel on each BeadChip, significantly reducing experimental variation and handling.

1.2. The WG-DASL Assay for Gene Expression Profiling

Formalin-fixed, paraffin-embedded (FFPE) tissues represent an invaluable resource for cancer research, as they are the most widely available material for which patient outcomes are known. There were over 300 million archived cancer tissue samples in the USA in 1999, with more samples accumulating at a rate of over 20 million per year (2). The ability to perform gene expression profiling in these samples enables prospective studies as well as retrospective analyses in which expression profiles derived from archived material may be correlated with known clinical outcomes (3). The combined adverse biochemical effects of tissue handling/storage and formalin fixation on RNA have been well-documented (4, 5), yielding cross-linked, chemically modified, and degraded fragments (up to 50% of which may not contain intact poly-A tails) (6). To overcome the technical limitations associated with applying conventional microarray-based technologies to the analyses of FFPE samples, we developed a sensitive and reproducible gene expression profiling assay, DASL (cDNA-mediated annealing, selection, extension, and ligation), for parallel analysis of hundreds of genes (7) or the whole-genome (8) with highly degraded RNA samples. The DASL technology incorporates several key features that make it particularly well-adapted to robustly profile degraded RNA samples (Fig. 1). Firstly, the incorporation of random priming during cDNA synthesis circumvents the sole dependence on poly-A/oligo-dT-based priming. Secondly, a relatively short target sequence of only ~50 nucleotides is required for query oligonucleotide annealing, thus enabling effective quantification of short, degraded mRNA fragments. Thirdly, only the first-strand cDNA is generated, thereby minimizing variation that arises during the cDNA synthesis. Fourthly, a single set of universal PCR primers is used to uniformly amplify all targets, generating amplicons of ~100 bp,

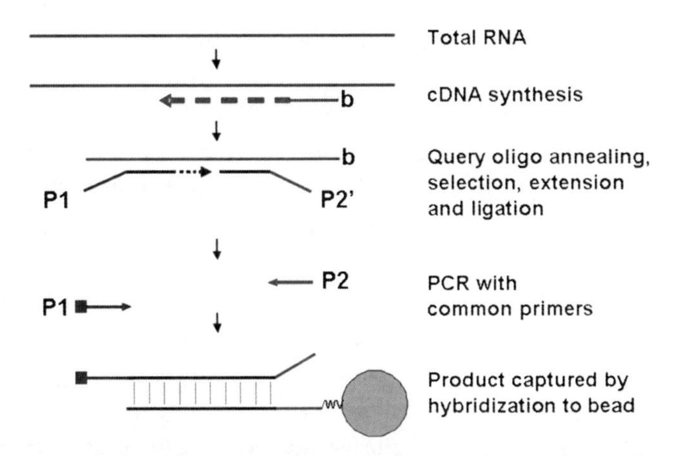

Fig. 1. WG-DASL assay scheme. In the WG-DASL assay, total RNA is converted to cDNA by random and oligo-dT priming. Two oligonucleotides are designed to uniquely target each of 24 K RefSeq transcripts, after which they are extended and ligated enzymatically. The ligated products are then amplified and fluorescently labeled during PCR, and finally detected by gene-specific hybridization to complementary 50-mer probe sequences on whole-genome gene expression BeadChips.

resulting in a relatively unbiased amplification of the PCR template population (7). Overall, the DASL technology combines the advantages of array-based gene expression analysis with those of multiplexed universal PCR, thereby affording much higher multiplexing capacity and throughput at lower cost. This opens up new avenues for large-scale discovery, validation, and clinical applications of mRNA biomarkers in human diseases.

2. Materials

Subheadings 2 and 3 assume that the user has access to either an Illumina iScan System or BeadArray Reader, and either the Universal Starter Kit or both the GoldenGate and Gene Expression Option kits and associated reagents and equipment.

2.1. Recommended Kit for RNA Extraction from FFPE Tissues

Of the few RNA preparation kits we tested, the High Pure RNA Paraffin Kit from Roche Applied Science (catalog # 03 270 289 001) yielded the highest quality RNA from FFPE samples for use in the WG-DASL assay (see Notes 1–3). Recently, more RNA extraction kits have become available (9).

2.2. Illumina-Supplied Reagents for the WG-DASL Assay

1. *M*aster Mix *c*DNA synthesis for *S*ingle Use *3* (MCS3), reagent for cDNA synthesis.
2. Whole-Genome *D*ASL *A*ssay *P*ool (DAP), mixture of oligonucleotides designed to query cDNA target sequences.

3. *O*ligo-*b*inding buffer *1* (OB1), oligo-annealing buffer which also contains paramagnetic particles to optimize washing, extension, and ligation steps of assay.
4. *A*dd *M*EL buffer 1 (AM1), wash buffer used to remove excess mis- or unhybridized query oligonucleotides.
5. *U*niversal wash *B*uffer 1 (UB1), wash buffer for several pre-PCR steps.
6. *M*aster mix for *E*xtension and *L*igation (MEL), optimized mixture of enzymes for extension/ligation step.
7. *S*ingle *C*olor *M*aster mix (SCM), PCR master mix which contains fluorescent and biotinylated common primers for multiplexed ligated oligonucleotide templates.
8. *I*noculate *P*CR buffer 1 (IP1), elution buffer for inoculating PCR reaction with ligated templates.
9. *M*agnetic *P*article buffer B (MPB), suspension of paramagnetic particles used to bind PCR products.
10. *U*niversal wash *B*uffer 2 (UB2), wash buffer used in post-PCR process.
11. *M*ake *H*yb buffer 1 (MH1), buffer used to neutralize and prepare single-stranded sample for hybridization to BeadChip.
12. *P*recipitation *S*olution 1 (PS1), reagent used to precipitate single-stranded DNA.
13. *Hy*bridization *B*uffer (HYB), buffer used in hybridization of sample to BeadChip.
14. *H*umidity *C*ontrol *B*uffer (HCB), buffer used to maintain humid environment in Hybridization (Hyb) Chamber.
15. *B*ead*C*hip solution (E1BC), wash buffer used in posthybridization steps.
16. *H*igh *T*emperature *W*ash buffer (HTW), wash buffer used during posthybridization.
17. *P*repare *B*eadChips solution *1* (PB1), buffer used during staining process.
18. *X*Stain Bead*C*hips solution *4* (XC4), solution used to coat BeadChips.

Please refer to Table 1 for reagent storage conditions.

2.3. Other Reagents Required for the Assay

1. 0.1 N NaOH (Sigma-Aldrich, S0899).
2. TE buffer: 10 mM Tris–HCl, 1 mM EDTA, pH 8.0.
3. DNA polymerase (Clontech, 639220).
4. Uracil DNA glycosylase (UDG), 1 U/μL (Invitrogen, 18054-015).
5. 2-Propanol.
6. 70 and 100% EtOH (ethanol).

Table 1
Reagent storage conditions

Reagent	Storage conditions	Shelf life (year)	Comments
MCS3	-25°C to -15°C	1	Aliquot to refreeze
DAP	-25°C to -15°C	2	Can be stored at 2–8°C up to 2 weeks
OB1	-25°C to -15°C	1	Does not completely freeze
AM1	2–8°C	1	
UB1	2–8°C	1	
MEL	-25°C to -15°C	1	Aliquot to refreeze
SCM	-25°C to -15°C	1	Aliquot to refreeze after adding DNA polymerase
IP1	-25°C to -15°C	1	
MPB	2–8°C	1	Do not freeze
UB2	RT	1	
MH1	RT	1	Keep away from light
PS1	2–8°C	1	
HYB	-25°C to -15°C	1	Keep away from light
HCB	-25°C to -15°C	1	
E1BC	RT	1	
HTW	RT	1	
PB1	RT	1	
XC4	RT	1	

7. Quant-iT RiboGreen RNA Assay Kit (Invitrogen, R-11490).
8. Serological pipettes (10, 25, and 50 mL).
9. 96-well, 0.2-mL skirted microtiter plates (Bio-Rad, MSP-9601).
10. 96-well, black, flat-bottom Fluotrac 200 plates (Greiner Bio One, 655076).
11. 96-well cap mats, sealing mats, round cap, pierceable, nonautoclavable (ABgene, AB-0566).
12. Heat sealing foil sheets, thermo seal (ABgene, AB-0559).
13. Microtiter plate clear adhesive film, 2mil Sealplate Adhesive Film, Nonsterile (Phenix Research Products, LMT-SEAL-EX).
14. Microseal “A” Film, PCR plate sealing film (Bio-Rad, MSA-5001).
15. Microseal “F” Film, aluminum adhesive film (Bio-Rad, MSF-1001).

16. 96-well V-bottom plates, Corning Costar Polypropylene (Fisher Scientific, 07-200-695 or VWR, 29444-102).
17. Multiscreen Filter plates, 0.45 µm, clear, Styrene (Millipore, MAHV-N45 10/50).
18. Multiscreen centrifuge alignment frames (Millipore, MACF09604).
19. Nonsterile solution basins (Labcor Products Inc., 730-001 or VWR, 21007-970).

2.4. Array Imaging

The Illumina iScan System or BeadArray Reader is required for array imaging. Loading and unloading of BeadChips into the iScan System or BeadArray Reader can be automated with the optional AutoLoader2 or AutoLoader, respectively. Both AutoLoaders support unattended processing by placing BeadChips carriers in the imaging system's tray so that it can scan the BeadChips. Both scanners use a laser to excite the fluor of the hybridized single-stranded product on the beads of the BeadChip sections. Light emissions from these fluors are then recorded in high-resolution images of the BeadChip sections. Data from these images are analyzed using Illumina's GenomeStudio Gene Expression Module. For Illumina iScan System or BeadArray Reader instructions, see the respective User Guides (http://www.illumina.com/support/documentation.ilmn).

3. Methods

3.1. RNA Extraction from FFPE Tissues

RNA extraction from FFPE samples is a critical step for success in the WG-DASL assay. We recommend the High Pure RNA Paraffin Kit (Roche) for RNA purification, using five 5-µm sections of FFPE tissues. All buffers and reagents are provided with the kit. The RNA extraction is carried out following the manufacturers' recommendations (see Notes 1 and 2). RNA integrity can be assessed with the Agilent 2100 Bioanalyzer using the RNA 6000 Pico Chip Kit.

3.2. WG-DASL Assay Protocols

These protocols are not intended to replace the WG-DASL assay guide supplied with Illumina systems, but rather give a detailed overview of the process. Please consult the WG-DASL assay guide (http://www.illumina.com/support/documentation.ilmn) for the latest protocol updates before performing any experiments.

3.2.1. Assay Probe Design

The WG-DASL assay is based on the sequence-specific extension and ligation of correctly hybridized query oligonucleotides, which are distinguished by their shared primer landing sites (for an overview of the WG-DASL assay workflow, see http://www.illumina.com/

support/literature.ilmn). Query oligonucleotides were designed to target intraexonic sequences and nonpolymorphic sites in cDNAs, using bioinformatic and biochemical parameters optimized for the assay biochemistry (see Notes 4 and 5). The assay uses two assay-specific oligonucleotides to interrogate a single unique contiguous 50 nt sequence on each cDNA (Fig. 1). Each of these oligonucleotides consists of two parts: an upstream-specific oligonucleotide (USO) containing a 3′ gene-specific sequence and a 5′ universal PCR primer sequence (P1) while the downstream-specific oligonucleotide (DSO) contains a 5′ gene-specific sequence and a different 3′ universal PCR primer sequence (P2′). Using this approach, a total of 24,526 oligonucleotide pairs (probes) were designed and pooled, which together constitutes the WG-DASL Assay Pool (DAP), corresponding to 18,626 unique genes, based on well-annotated content derived from the National Center for Biotechnology Information Reference Sequence Database (Build 36.2, Release 22). The DAP is then annealed to the targeted cDNAs followed by enzymatic extension and ligation steps. Ligated products are PCR-amplified and labeled with a pair of Cy3-coupled and biotinylated universal primers, after which single-stranded labeled products are hybridized to whole-genome gene expression BeadChips bearing complementary gene-specific probe sequences (8).

3.2.2. WG-DASL Assay Protocol

The Make *S*ingle *U*se *R*NA Process for cDNA Synthesis

1. Preheat heat block to 42°C and allow temperature to stabilize. Thaw cDNA synthesis reagent (MCS3) to room temperature and vortex to fully mix tube contents.
2. Normalize partially degraded RNA samples to 40–200 ng/μL or intact RNA samples to 20–100 ng/μL with DEPC-treated H_2O. We recommend using 200 ng of total RNA extracted from FFPE samples (see Notes 6 and 7) and 100 ng of total RNA extracted from intact or fresh-frozen samples (see Note 8) for one assay.
3. Add 5 μL of MCS3 to each well of microtiter plate labeled *S*ingle *U*se *R*NA (SUR). Transfer 5 μL of normalized RNA sample to each well of SUR plate, heat seal with foil, and vortex at 2,300 rpm for 20 s. Pulse-centrifuge sealed plate to 250 × *g* to prevent evaporation from wells during incubation.
4. Incubate SUR plate at room temperature for at least 10 min and then continue incubation at 42°C for 60 min in preheated heat block.
5. When incubation is complete, remove SUR plate from heat block and pulse-centrifuge to 250 × *g* to collect any condensation.
6. If experiment is continued the same day, set heat block to 70°C. If not proceeding with experiment immediately, SUR plate can be stored up to 4 h at 2–8°C or overnight at −15°C to −25°C.

The Make *A*ssay-*S*pecific *E*xtension Process for Annealing of Query Oligonucleotides to cDNA

1. Preheat heat block to 70°C and allow temperature to stabilize.
2. Remove DAP tube from freezer (if frozen, thaw, vortex, and then centrifuge). Thaw Oligo-binding buffer 1 (OB1) to room temperature and vortex. Do not centrifuge OB1 tube.
3. Dispense 10 μL of DAP to each well of a new, 96-well, 0.2-mL skirted microtiter plate labeled "ASE."
4. Add 30 μL of well-resuspended OB1 to each well of *a*ssay-*s*pecific *e*xtension (ASE) plate.
5. Centrifuge SUR plate to 250 × *g* to collect samples at bottom of wells. Transfer 10 μL of biotinylated cDNA to ASE plate containing DAP and OB1 to bring final volume to 50 μL. Heat seal ASE plate and vortex briefly at 1,600 rpm to mix content of wells. Place ASE plate in 70°C heat block and immediately reduce temperature setting to 30°C. This allows oligonucleotide annealing to cDNA targets by ramping temperature over approximately 2 h then holding at 30°C until next processing step.

The Add MEL Process for Assay Oligonucleotide Extension and Ligation

1. Remove ASE plate from heat block, reset it to 45°C and allow the temperature to stabilize. Thaw MEL reagent to room temperature.
2. Place ASE plate with oligonucleotides annealed to cDNA templates on Illumina-supplied magnetic plate for at least 2 min, or until beads are completely captured. Washing beads remove excess and mishybridized oligonucleotides.
3. After paramagnetic particles are captured, remove heat seal from plate and remove and discard all liquid (~50 μL) from wells, retaining beads. Add 50 μL AM1 to each well of assay plate. Seal plate with adhesive film and vortex at 1,600 rpm for 20 s, or until all beads are resuspended.
4. Place ASE plate on magnet for at least 2 min, or until beads are completely captured. Remove all AM1 from each well, leaving beads in wells. Repeat addition of 50 μL AM1, vortexing, and removal of buffer.
5. Remove ASE plate from magnet and add 50 μL of UB1 to each well.
6. Place ASE plate on magnet for at least 2 min, or until beads are completely captured. Remove all UB1 from each well. Repeat addition of 50 μL UB1 and removal of buffer.
7. Add 37 μL of MEL to each well of ASE plate. Seal plate with adhesive film and vortex at 1,600 rpm for 1 min.
8. Incubate ASE plate on preheated 45°C heat block for 15 min.

The Make PCR and Inoculate PCR Processes for Preparing the PCR Mix and Setting up the PCR Reaction

1. Prepare PCR master mix by adding 64 μL of DNA polymerase and 50 μL UDG to tube of SCM PCR reagent. Invert SCM tube several times to mix contents and aliquot 30 μL of mixture into each well of new 96-well, 0.2-mL microtiter (PCR) plate.
2. Remove ASE plate from heat block after extension and ligation step and reset heat block to 95°C.
3. Place ASE plate on magnet for at least 2 min, or until beads are captured. Remove clear adhesive film from assay plate, and remove and discard supernatant (~50 μL) from all wells of ASE plate, leaving beads in wells. Leave ASE plate on magnet and add 50 μL UB1 to each well of plate.
4. Allow ASE plate to rest on magnet for at least 2 min to collect paramagnetic particles. Remove and discard all supernatant (~50 μL) from all wells of ASE plate, leaving beads in wells.
5. Add 35 μL of IP1 to each well of assay plate and seal it with adhesive film. Vortex plate at 1,800 rpm for 1 min, or until all beads are resuspended. Place plate on 95°C heat block for 1 min.
6. Remove ASE plate from heat block and place it on magnet for at least 2 min, or until beads have been completely captured. Transfer 30 μL of supernatant from first column of ASE plate into first column of PCR plate. Repeat transfer for remaining columns, using new pipette tips for each column.
7. Seal PCR plate with microseal "A" PCR plate sealing film.
8. Pulse-centrifuge PCR plate to 250 × *g* for 1 min. Immediately transfer PCR plate to thermal cycler and run following cycling program: 10 min at 37°C; 3 min at 95°C; 34 cycles (35 s at 95°C, 35 s at 56°C, 2 min at 72°C); 10 min at 72°C; and 4°C for 5 min.
9. Proceed immediately to preparation of single-stranded PCR products for precipitation or seal and store PCR plate at −15°C to −25°C.

The Make Intermediate Plate Process to Prepare Samples for Precipitation

1. Vortex magnetic particle buffer B reagent (MPB) until beads are completely resuspended.
2. Dispense 20 μL of resuspended MPB into each well of PCR plate. Mix beads with PCR product by pipetting up and down, and then transfer mixed solution to filter plate. Cover filter plate with its cover and store at room temperature, protected from light, for 60 min.
3. Place filter plate containing bound PCR products onto new 96-well V-bottom waste plate using filter plate adapter. Centrifuge at 1,000 × *g* for 5 min at 25°C.

4. Remove filter plate lid. Add 50 μL of UB2 to each well of filter plate. Dispense slowly so that beads are undisturbed. Replace lid of filter plate and centrifuge at 1,000 × *g* for 5 min at 25°C.
5. Prepare new 96-well, V-bottom plate, and dispense 30 μL of MH1 to all wells of new intermediate plate. Place filter plate onto intermediate plate such that column A1 of filter plate matches column A1 of intermediate plate.
6. Dispense 30 μL of 0.1 N NaOH to all wells of filter plate. Replace lid of filter plate and centrifuge immediately at 1,000 × *g* for 5 min at 25°C. Gently mix contents of intermediate plate by moving it from side to side, without splashing.
7. Seal INT plate with 96-well cap mat. If precipitation is not performed immediately, INT plate can be stored at −15°C to −25°C for up to 24 h.

The Precipitate and Wash Intermediate Plate for the BeadChip Process to Prepare Samples for Array Hybridization

Preheat hybridization oven to 58°C and allow temperature to equilibrate. Place HYB tube in 58°C oven for 10 min to dissolve any salts that may have precipitated in storage. If any salts remain undissolved, incubate at 58°C for another 10 min. Cool to room temperature and mix thoroughly before using.

1. If INT plate has been frozen, thaw it completely at room temperature in light-protected drawer.
2. Preheat heat block to 65°C and allow temperature to stabilize.
3. Prepare MH1/water/HYB mix by combining 300 μL MH1, 300 μL water, and 1.2 mL HYB. Mix well by vortexing. Store in light-protected drawer.
4. Add 30 μL PS1 to each well of INT plate. Using multichannel pipette, thoroughly mix contents by pipetting solution up and down several times until solution is uniformly blue.
5. Add 90 μL 2-propanol to each well of INT plate. Using multichannel pipette, thoroughly mix contents by pipetting solution up and down several times until solution is uniformly blue to ensure efficient precipitation.
6. Seal INT plate with clear adhesive film and centrifuge plate to 3,000 × *g* at 2–8°C for 20 min.
7. Remove INT plate seal and decant supernatant by inverting INT plate and smacking it down onto absorbent pad. Tap inverted plate onto pad to blot excess supernatant.
8. Add 150 μL 70% ethanol (EtOH) to each well of INT plate. Using multichannel pipette, thoroughly wash blue pellet in 70% EtOH by pipetting up and down several times.
9. Seal INT plate with clear adhesive film and centrifuge plate to 3,000 × *g* at 2–8°C for 10 min.

10. Remove INT plate seal and decant supernatant by inverting INT plate and smacking it down onto absorbent pad. Tap inverted plate onto pad to blot excess supernatant.
11. Place INT plate at 65°C in preheated heat block for 5 min, or until residual EtOH has evaporated.
12. Add 15 μL of MH1/water/HYB mix to each well of INT plate.
13. Seal INT plate with clear adhesive film and pulse-centrifuge to 250 × *g*.
14. Remove INT plate seal and using multichannel pipette, thoroughly dissolve blue pellets by pipetting solution up and down several times.
15. Seal plate with 96-well cap mat. If hybridization is not performed immediately, INT plate can be stored at −15°C to −25°C for up to 24 h.

3.3. Hybridize to BeadChip

1. Preheat hybridization oven to 58°C and allow temperature to equilibrate. Place humidity control buffer (HCB) tube in 58°C oven for 10 min to dissolve any salts that may have precipitated in storage. If any salts remain undissolved, incubate at 58°C for another 10 min. Cool to room temperature and mix thoroughly before using.
2. If INT plate has been frozen, thaw it completely at room temperature in light-protected drawer. Pulse-centrifuge plate at 250 × *g* for 1 min.
3. Prepare hybridization (Hyb) chambers by pipetting 200 μL HCB into Hyb chamber reservoirs, followed by sealing Hyb chambers as described in WG-DASL assay guide.
4. Place multisample BeadChips into Hyb chamber inserts.
5. Manually pipette 15 μL of sample into appropriate inlet ports on multisample BeadChips.
6. Place inserts with sample-laden BeadChips into Hyb chamber, seal, and hybridize samples, with rocking, at 58°C for 16–20 h.
7. In preparation for next day's washes, prepare 1× HTW buffer from 10× stock by adding 50 mL 10× HTW buffer to 450 mL nuclease-free water.
8. Place Hybex waterbath insert into Hybex heating base.
9. Add 500 mL prepared 1× HTW buffer into Hybex waterbath insert.
10. Set Hybex heating base temperature to 55°C.
11. Close Hybex heating base lid and leave HTW buffer to warm overnight.
12. Proceed to Wash BeadChip step the next day.

3.4. Wash BeadChip

1. In preparation for Coat BeadChip step, remove XStain BeadChip solution 4 (XC4) reagent bottle from the freezer and thaw.
2. Add 335 mL 100% EtOH to XC4 bottle. The final volume will be 350 mL.
3. Each XC4 bottle contains enough to process up to 24 BeadChips.
4. Recap bottle, shake vigorously for 15 s, and place on rocker for 30–40 min to resuspend. Place bottle on side opposite to frozen pellet if possible.
5. After 30–40 min, shake bottle vigorously by hand to ensure that all XC4 is in suspension, and none is still coating the container. If coating is visible, vortex at 1,625 rpm until XC4 is in complete suspension.
6. Once resuspended, use XC4 at room temperature. You can store it at 2–8°C overnight. Keep XC4 in bottle in which it was shipped until ready for use.
7. Add 6 mL E1BC buffer to 2 L RNase-free water to make Wash E1BC solution. Place 1 L of diluted Wash E1BC buffer in Pyrex No. 3140 beaker.
8. Remove Hyb chamber from hybridization oven and place it on lab bench. Disassemble Hyb chamber.
9. Using powder-free gloved hands, remove all BeadChips from Hyb chamber and submerge them face up at bottom of beaker.
10. Using powder-free gloved hands, remove coverseal from first BeadChip. Ensure that entire BeadChip remains submerged during removal.
11. Using tweezers or powder-free gloved hands, transfer BeadChip to slide rack submerged in dish containing 250 mL Wash E1BC solution.
12. Repeat steps 10 and 11 for all BeadChips from same Hyb chamber.
13. Using slide rack handle, transfer rack into Hybex waterbath insert containing preheated HTW buffer.
14. Incubate static for 10 min with Hybex lid closed.
15. After 10 min incubation in HTW buffer is complete, immediately transfer slide rack back into dish containing 250 mL fresh Wash E1BC buffer.
16. Using slide rack handle, plunge rack in and out of solution 5–10 times.
17. Place dish on orbital shaker and shake at medium-low speed at room temperature for 5 min.

18. Transfer rack to clean dish containing 250 mL fresh 100% EtOH.
19. Using slide rack handle, plunge rack in and out of solution 5–10 times.
20. Place dish on orbital shaker and shake at medium-low speed at room temperature for 10 min.

3.5. Coat BeadChip

1. Fill vertical wash dish with 310 mL of Prepare BeadChip solution 1 (PB1).
2. Submerge unloaded staining rack into wash dish with locking arms and tab facing you.
3. Quickly transfer each BeadChip from EtOH wash to staining rack while it is submerged in PB1. BeadChip barcodes should face away from you (do not let BeadChips dry).
4. Wash BeadChips in PB1 by moving staining rack up and down ten times (break surface each time).
5. Soak BeadChips for 5 min in PB1.
6. Fill vertical wash dish with 310 mL of XC4 (should be used within 10 min).
7. Remove BeadChips in staining rack from PB1 wash dish and place in XC4 wash dish.
8. Wash in XC4 with physical motion described in step 4 followed by a soak for 5 min.
9. Remove staining rack in one smooth motion going from a vertical orientation to a horizontal orientation with barcodes on BeadChips facing upward. Place on tube rack.
10. Individually remove BeadChips from staining rack and place barcode side up on tube rack making sure BeadChips are not touching each other during drying step.
11. Place tube rack with horizontal BeadChips into vacuum desiccator, turn on vacuum to ~508 mmHg, and dry for 50–55 min at room temperature.

3.6. Image BeadChip on iScan System or BeadArray Reader

The arrays are imaged using either the iScan System or the BeadArray Reader. Image processing and intensity data extraction are performed by the iScan Control Software (or BeadScan, for the BeadArray Reader). The iScan System incorporates advanced optics and sensors to support much higher throughput than the BeadArray Reader while providing equally high data quality.

3.7. Data Collection and Analysis

The GenomeStudio software package is included with the WG-DASL assay product, and is used as a tool for analyzing gene expression data from scanned microarray images collected from either the iScan System or the BeadArray Reader. Alternatively,

GenomeStudio can be used to export the array intensity data for processing by most standard gene expression analysis programs. Specifically, GenomeStudio executes two types of data analysis:

1. Gene analysis, quantifying gene expression levels.
2. Differential analysis, determining whether gene expression levels are different between two experimental groups.

Analyses can be performed on individual samples or groups of samples.

GenomeStudio reports experiment performance based on built-in controls that accompany each experiment (see Notes 9–12). In addition, GenomeStudio provides plotting (line plots, box plots, and scatter plots) tools, is able to generate dendrograms and heatmaps, and includes a genome viewer and chromosome browser all of which facilitate quick and visual means for exploratory analyses (see Note 13).

3.8. Examples of WG-DASL Assay Applications

DASL technology has been successfully used to profile a variety of archived FFPE tumor samples, some of which have been in storage for as long as 24 years (10) and for which, in many cases, little or no tissue handling and fixation details are known, including colon (11, 12), breast (9, 11–15), lung (12, 13), prostate (10, 12, 16–18), bladder (15), and liver (19) cancer, as well as autopsied tissue (13). More recently, the WG-DASL assay has been used successfully to identify known as well as discover new markers associated with high-grade stage I serous epithelial ovarian cancer (20).

4. Notes

1. RNA from FFPE specimens can be difficult to extract, since the RNA becomes cross-linked and degraded during the fixation and storage process; in addition, the amount of tissue in the FFPE specimen can be very small (e.g., normal lung samples consist mostly of extracellular air space). Therefore, it is essential to have a robust method to efficiently retrieve high-quality RNA from FFPE tissue. In our experience, the Roche High Pure RNA Paraffin Kit results in superior quality RNA as compared to other commercially available kits for RNA preparation from FFPE samples (11); other groups have reported the same result (9). However, this protocol is tedious and time consuming, which may pose a limitation for high sample throughput.
2. We recommend using several 5-μm sections of FFPE tissue for RNA extraction (21, 22). Our results suggest that 5-μm tissue sections yielded more RNA than thicker 20-μm sections,

probably due to more efficient deparaffinization and cell lysis. The average tissue sections from paraffin blocks were approximately 1 cm^2 with the solid tissue in the middle. We also isolated RNA from tissue sections mounted on slides.

3. RNA quantitation is an important step to ensure that sufficient material is used in the WG-DASL assay to generate high-quality data. We recommend the Quant-iT RiboGreen RNA Assay Kit (Invitrogen) for quantitation of RNA samples. The RiboGreen assay measures RNA directly and can quantitate small RNA volumes. Other quantitation methods may be prone to measuring contaminants, such as small molecules and proteins. We also recommend using a fluorometer, as fluorometry provides RNA-specific quantitation, whereas spectrophotometry may be affected by DNA contamination, leading to artificially inflated amounts.
4. Careful experimental design may help maximize the utility of the WG-DASL assay. Because the DAP oligonucleotides target intraexonic sequences, genomic DNA samples can be used to monitor assay performance and troubleshoot questionable WG-DASL assay results for FFPE-derived RNA samples. As an option, one can use the make single-use DNA (Make SUD) process to prepare activated gDNA to include as a positive control sample. The data derived from the gDNA samples can then be used to qualify individual target assays during data analysis in the GenomeStudio application. A well-designed WG-DASL assay experiment includes replicates and positive control samples, such as gDNA and intact reference RNA. We find that the gDNA samples are useful to monitor assay performance in the steps following cDNA synthesis, and the reference RNA sample can be used as a positive control relative to the degraded FFPE-derived RNA. Replicates can identify those samples that are less robust and less trustworthy by lower correlations.
5. As mentioned previously, a gDNA sample can be used to qualify the performance of the assay as well as the arrays. Ideally, intensities for all probes should exceed background for a gDNA sample. Exceptions occur when individual probes are not uniquely mapped to the genome in RefSeq, causing a probe to cross an exon junction.
6. The data quality for samples with too little input is quite similar to those with too much RNA degradation. We observed that poor samples are characterized by a decrease in overall signal across probes, a shift in the distribution of signals to lower intensities (i.e., fewer genes with high signals), an increase in the intensity of the negative controls, and an increase in the variation among sample replicates. To estimate the impact of input RNA quantity on assay performance, we tested various

amounts of total FFPE-derived RNA (200 ng, 50 ng, 10 ng, 5 ng, 1 ng, and 50 pg) in the WG-DASL assay (8). Each cDNA sample was processed in duplicate for independent technical replicates. FFPE RNA inputs as low as 5 ng yielded detected probe concordance (overlap) rates of >85% and raw intensity correlations of $R^2 > 0.85$ when compared with standard inputs of 200 ng (Table 2). Moreover, these results were reproducible with self-correlations of $R^2 \sim 0.90$. Despite the lower probe concordance rates (~60%) between the 1 ng input and the standard 200 ng input, the raw expression profiles at the 1 ng input level remained reproducible, with $R^2 \sim 0.80$ and had correlations of $R^2 > 0.75$ when compared with the standard 200 ng input.

7. We previously assessed the performance of the WG-DASL assay in matched pairs of fresh-frozen (FF) and FFPE samples, using RNAs derived from normal adjacent tissue (NAT) and tumor (TUM) lung tissue (8). We obtained a high degree of sensitivity, with the number of detected transcripts in both the NAT and TUM FFPE samples approximately 90% of that detected in the corresponding FF samples. Moreover, the overlap of the detected transcripts was also high, with 94.2% (14188/15064; NAT) and 96.3% (14066/14610; TUM) of the FFPE transcripts also detected in the corresponding FF sample. Direct intensity comparisons between matched FF and FFPE samples yielded average correlations of $R^2 \sim 0.62$; fold-change comparisons of several thousand detected transcripts between paired FF and FFPE samples were higher (average $R^2 \sim 0.70$), with a slope approximating 45° and no significant compression or expansion observed for the log fold-change ratios. We also generated lists of differentially expressed transcripts between the TUM and NAT samples for both FF and FFPE tissues, and then determined the extent of the overlap between the two differentially expressed transcript lists. With a false discovery rate (FDR) of <5%, approximately 74% (4793/6473) of the transcripts identified in the FFPE comparison also overlapped with those identified in the FF comparison. Similar overlapping results were obtained at a more stringent FDR (<1%). These results suggest that sets of differentially expressed genes identified in FFPE samples are similar to those identified in FF samples. The correlation (R^2) between the gene expression profiles of paired FFPE and FF tissues was 0.62 on average. This observation is entirely consistent with other measurements of expression results in FFPE versus FF tissues, including qPCR data. The result suggests that the large difference between the profiles generated from the matched FFPE and FF samples may be derived largely from variations in sample processing and handling, rather than the WG-DASL

Table 2
WG-DASL assay performance as a function of FFPE RNA input

Metric	200 ng	50 ng	10 ng	5 ng	1 ng	250 pg
Average number of detected probes (DFCI #120)	19,066	18,393	18,152	17,627	12,742	6,079
Average number of detected probes (DFCI #129)	18,031	17,677	17,113	16,404	10,998	5,585
Average probe overlap with 200 ng (DFCI #120; %)	100.0	93.6	91.0	88.7	63.7	30.8
Average probe overlap with 200 ng (DFCI #129; %)	100.0	95.7	90.8	87.3	60.5	30.8
Average correlation with 200 ng (DFCI #120; R^2)	100.0	0.951	0.920	0.895	0.794	0.636
Average correlation with 200 ng (DFCI #129; R^2)	100.0	0.957	0.903	0.873	0.748	0.398
Self-reproducibility (DFCI #120; R^2)	0.987	0.965	0.959	0.935	0.800	0.724
Self-reproducibility (DFCI #129; R^2)	0.986	0.979	0.958	0.940	0.803	0.680

Values for the number of detected probes, probe overlap, and correlation are derived from the average of two technical replicates. Probe overlap is calculated as a percentage of the number of probes with matching detected calls at p-value < 0.01 between the low and standard inputs divided by the total number of probes detected in the standard input (200 ng). Values shown for the correlation and self-reproducibility are calculated for probes detected ($p < 0.01$) in both inputs and replicates, respectively assay. It is well known that the quality of RNA extracted from FFPE samples depends largely on multiple parameters: initial tissue fixation, sample processing, storage, and the RNA extraction protocol. We believe that the major source of variability in expression profiles using RNA extracted from FFPE tissues is sample quality. Using a qPCR-based sample prequalification approach, we previously showed that FFPE samples with the lowest correlations with matching FF samples also had the poorest overall quality (8, 11). In addition, RNA secondary structure may change after degradation, which may lead to changes in random primer accessibility and annealing efficiency in the cDNA synthesis. Taken together, these observations suggest that sequence-dependent effects may be responsible for part of the variation in expression data between FF and FFPE samples. While we realize that more experiments comparing FF and matching FFPE samples are needed to study all possible sources and mechanisms of variation in gene expression between frozen and archived tissues, we strongly believe that FFPE samples maintain most of the biological signatures/information from the original tissues, and they can be used as an important resource for biomarker discovery and validation, particularly if comparisons are made between equivalently treated samples.

8. To determine the performance of the WG-DASL assay at low intact RNA input levels, we assayed various amounts (100 ng, 50 ng, 10 ng, 5 ng, 1 ng, 250 pg, 100 pg, and 10 pg) of Raji and MCF-7 total RNAs (8). For both cell lines, data generated with as little as 250 pg still yielded reproducible expression profiles with an average correlation of $R^2 \sim 0.92$ (Table 3), while the 50 pg and 10 pg RNA inputs yielded average intensity correlations of $R^2 \sim 0.82$ and 0.78, respectively. On average, the probe overlap as compared to 100 ng was 71.7%, 42.6%, and 22.9% for the 250, 50 and 10 pg inputs, respectively. Comparison of expression profiles generated with the lower inputs to that obtained with the standard 100 ng input revealed average intensity correlations of $R^2 \sim 0.91$ for the 250 pg, and $R^2 \sim 0.80$ and 0.72 for the 50 pg and 10 pg inputs, respectively. To further assess the ability of the WG-DASL assay to detect differentially expressed genes with lower inputs, we compared lists of differentially expressed transcripts detected at the 250 pg level to that obtained with the standard 100 ng input. Using a p-value = 0.001 cutoff, approximately 95.5% (10716/11222 probes) of the differentially expressed transcripts in the 250 pg input was also identified in the standard 100 ng input. We also compared the WG-DASL assay results to that obtained with qPCR. Here, we designed a set of primers corresponding to a panel of 24 genes, whose expression levels differed between the assayed cell lines (Raji and MCF-7).

Table 3
WG-DASL assay performance as a function of intact RNA input

Metric	100 ng	50 ng	10 ng	5 ng	1 ng	250 pg	50 pg	10 pg
Average number of detected probes (MCF-7)	15,247	15,150	15,060	14,966	14,279	10,382	6,671	3,258
Average number of detected probes (Raji)	13,291	13,141	13,032	13,001	12,197	10,001	5,499	3,254
Average probe overlap with 100 ng (MCF-7; %)	100.0	99.4	98.8	98.2	93.7	68.1	43.8	21.4
Average probe overlap with 100 ng (Raji; %)	100.0	98.9	98.1	97.8	91.8	75.2	41.4	24.5
Average correlation with 100 ng (MCF-7; R^2)	1.000	0.986	0.976	0.966	0.932	0.906	0.795	0.717
Average correlation with 100 ng (Raji; R^2)	1.000	0.988	0.976	0.971	0.920	0.917	0.807	0.729
Self-reproducibility (MCF-7; R^2)	0.988	0.987	0.986	0.982	0.969	0.909	0.826	0.791
Self-reproducibility (Raji; R^2)	0.990	0.989	0.987	0.986	0.941	0.921	0.821	0.775

Values for the number of detected probes, probe overlap, and correlation are derived from the average of two technical replicates. Probe overlap is calculated as a percentage of the number of probes with matching detected calls at p-value < 0.01 between the low and standard inputs divided by the total number of probes detected in the standard input (100 ng). Values shown for the correlation and self-reproducibility are calculated for probes detected ($p < 0.01$) in both inputs and replicates, respectively

Our comparisons across these 24 common genes between the WG-DASL (both 100 ng and 250 pg input RNA) and the qPCR assays consistently demonstrated, on average, a strong correlation (R^2 ~ 0.87 for the 100 ng and R^2 ~ 0.87 for the 250 pg inputs, respectively) across replicate experiments.

9. The annealing controls monitor the efficiency of annealing oligonucleotides with different melting temperatures to the same cDNA target. In each case, the higher melting temperature oligonucleotide probe should give higher signals than the lower melting temperature oligonucleotide.

10. The array hybridization controls test the hybridization of single-stranded assay products to the array beads. The controls consist of 50-mer oligonucleotides labeled with Cy3 dye included in the HYB reagent. Two types of controls comprise this category: the Cy3-labeled and the low stringency hybridization controls. The Cy3 hybridization controls consist of six probes with corresponding Cy3-labeled oligonucleotides present in the HYB reagent. Following successful hybridization, they produce a signal independent of both the cellular RNA quality and success of the sample preparation reactions. Target oligonucleotides for the Cy3 hybridization controls are present at three concentrations (low, medium, and high), yielding gradient hybridization responses. The low stringency hybridization control contains four probes, corresponding to the medium and high-concentration Cy3 hybridization control targets. In this case, each probe has two mismatch bases distributed in its sequence. If the hybridization stringency is adequate, these controls yield very low signals. If the stringency is too low, they yield signals approaching that of their perfect match counterparts in the Cy3 hybridization control category.

11. The negative controls consist of query oligonucleotides targeting ~300 random sequences that do not appear in the human genome. The mean signal of these probes defines the system background. This background is represented by both the imaging system background and by any signal resulting from cross-hybridization or nonspecific binding of dye. The GenomeStudio software platform uses the signals and signal standard deviation of these probes to establish gene expression detection limits. Any outlier samples that show high counts or large standard deviations in the negative controls should be considered with caution. We find that high negative controls can be associated with RNAs that are overly degraded, and therefore are unlikely to provide useful expression information.

12. The control summary graph titled "Genes" reports the average intensity and standard deviation for all genes. This value is expected to be lower in RNA samples than gDNA samples and lower in FFPE-derived RNAs than in intact RNAs.

Outlier samples with low intensities may reflect overly degraded RNA or very low RNA input. Because the assay monitors the number of amplifiable targets, low RNA input levels can resemble overly degraded RNAs in WG-DASL assay performance.

13. Another valuable approach to assessing data quality is to cluster the samples using the dendrogram tool in GenomeStudio. If there are replicate samples or known sample relationships, outlier samples can be readily identified by their failure to associate with replicate or related samples by hierarchical clustering.

Acknowledgments

We thank Marina Bibikova, Mark Staebell, and Brent Applegate at Illumina, Monica Reinholz and Jeremy Chien at Mayo Clinic for helpful discussions.

References

1. Barker, D. L., Theriault, G., Che, D., Dickinson, T., Shen, R., and Kain, R. (2003) Self-assembled random arrays: High-performance imaging and genomics applications on a high-density microarray platform *Proc SPIE* **4966**, 1–11.
2. Bouchie, A. (2004) Coming soon: a global grid for cancer research *Nat Biotechnol* **22**, 1071–3.
3. Ramaswamy, S. (2004) Translating cancer genomics into clinical oncology *N Engl J Med* **350**, 1814–6.
4. Medeiros, F., Rigl, C. T., Anderson, G. G., Becker, S. H., and Halling, K. C. (2007) Tissue handling for genome-wide expression analysis: a review of the issues, evidence, and opportunities *Arch Pathol Lab Med* **131**, 1805–16.
5. Hewitt, S. M., Lewis, F. A., Cao, Y., Conrad, R. C., Cronin, M., Danenberg, K. D., Goralski, T. J., Langmore, J. P., Raja, R. G., Williams, P. M., Palma, J. F., and Warrington, J. A. (2008) Tissue handling and specimen preparation in surgical pathology: issues concerning the recovery of nucleic acids from formalin-fixed, paraffin-embedded tissue *Arch Pathol Lab Med* **132**, 1929–35.
6. Masuda, N., Ohnishi, T., Kawamoto, S., Monden, M., and Okubo, K. (1999) Analysis of chemical modification of RNA from formalin-fixed samples and optimization of molecular biology applications for such samples *Nucleic Acids Res* **27**, 4436–43.
7. Fan, J. B., Yeakley, J. M., Bibikova, M., Chudin, E., Wickham, E., Chen, J., Doucet, D., Rigault, P., Zhang, B., Shen, R., McBride, C., Li, H. R., Fu, X. D., Oliphant, A., Barker, D. L., and Chee, M. S. (2004) A versatile assay for high-throughput gene expression profiling on universal array matrices *Genome Res* **14**, 878–85.
8. April, C., Klotzle, B., Royce, T., Wickham-Garcia, E., Boyaniwsky, T., Izzo, J., Cox, D., Jones, W., Rubio, R., Holton, K., Matulonis, U., Quackenbush, J., and Fan, J. B. (2009) Whole-genome gene expression profiling of formalin-fixed, paraffin-embedded tissue samples *PLoS One* **4**, e8162.
9. Abramovitz, M., Ordanic-Kodani, M., Wang, Y., Li, Z., Catzavelos, C., Bouzyk, M., Sledge, G. W., Jr., Moreno, C. S., and Leyland-Jones, B. (2008) Optimization of RNA extraction from FFPE tissues for expression profiling in the DASL assay *Biotechniques* **44**, 417–23.
10. Setlur, S. R., Mertz, K. D., Hoshida, Y., Demichelis, F., Lupien, M., Perner, S., Sboner, A., Pawitan, Y., Andren, O., Johnson, L. A., Tang, J., Adami, H. O., Calza, S., Chinnaiyan, A. M., Rhodes, D., Tomlins, S., Fall, K., Mucci, L. A., Kantoff, P. W., Stampfer, M. J., Andersson, S. O., Varenhorst, E., Johansson, J. E., Brown, M., Golub, T. R., and Rubin, M. A. (2008) Estrogen-dependent signaling in a molecularly distinct subclass of aggressive prostate cancer *J Natl Cancer Inst* **100**, 815–25.
11. Bibikova, M., Talantov, D., Chudin, E., Yeakley, J. M., Chen, J., Doucet, D., Wickham, E., Atkins, D., Barker, D., Chee, M., Wang, Y., and Fan, J. B. (2004) Quantitative gene expression profiling in formalin-fixed, paraffin-embedded

tissues using universal bead arrays *Am J Pathol* **165**, 1799–807.

12. Bibikova, M., Yeakley, J. M., Chudin, E., Chen, J., Wickham, E., Wang-Rodriguez, J., and Fan, J. B. (2004) Gene expression profiles in formalin-fixed, paraffin-embedded tissues obtained with a novel assay for microarray analysis *Clin Chem* **50**, 2384–6.
13. Haller, A. C., Kanakapalli, D., Walter, R., Alhasan, S., Eliason, J. F., and Everson, R. B. (2006) Transcriptional profiling of degraded RNA in cryopreserved and fixed tissue samples obtained at autopsy *BMC Clin Pathol* **6**, 9.
14. Paik, S. (2006) Methods for gene expression profiling in clinical trials of adjuvant breast cancer therapy *Clin Cancer Res* **12**, 1019 s–1023 s.
15. Ravo, M., Mutarelli, M., Ferraro, L., Grober, O. M., Paris, O., Tarallo, R., Vigilante, A., Cimino, D., De Bortoli, M., Nola, E., Cicatiello, L., and Weisz, A. (2008) Quantitative expression profiling of highly degraded RNA from formalin-fixed, paraffin-embedded breast tumor biopsies by oligonucleotide microarrays *Lab Invest* **88**, 430–40.
16. Bibikova, M., Chudin, E., Arsanjani, A., Zhou, L., Garcia, E. W., Modder, J., Kostelec, M., Barker, D., Downs, T., Fan, J. B., and Wang-Rodriguez, J. (2007) Expression signatures that correlated with Gleason score and relapse in prostate cancer *Genomics* **89**, 666–72.
17. Li, H. R., Wang-Rodriguez, J., Nair, T. M., Yeakley, J. M., Kwon, Y. S., Bibikova, M., Zheng, C., Zhou, L., Zhang, K., Downs, T., Fu, X. D., and Fan, J. B. (2006) Two-dimensional transcriptome profiling: identification of messenger RNA isoform signatures in prostate cancer from archived paraffin-embedded cancer specimens *Cancer Res* **66**, 4079–88.
18. Nakagawa, T., Kollmeyer, T. M., Morlan, B. W., Anderson, S. K., Bergstralh, E. J., Davis, B. J., Asmann, Y. W., Klee, G. G., Ballman, K. V., and Jenkins, R. B. (2008) A tissue biomarker panel predicting systemic progression after PSA recurrence post-definitive prostate cancer therapy *PLoS One* **3**, e2318.
19. Hoshida, Y., Villanueva, A., Kobayashi, M., Peix, J., Chiang, D. Y., Camargo, A., Gupta, S., Moore, J., Wrobel, M. J., Lerner, J., Reich, M., Chan, J. A., Glickman, J. N., Ikeda, K., Hashimoto, M., Watanabe, G., Daidone, M. G., Roayaie, S., Schwartz, M., Thung, S., Salvesen, H. B., Gabriel, S., Mazzaferro, V., Bruix, J., Friedman, S. L., Kumada, H., Llovet, J. M., and Golub, T. R. (2008) Gene expression in fixed tissues and outcome in hepatocellular carcinoma *N Engl J Med* **359**, 1995–2004.
20. Chien, J., Fan, J. B., Bell, D. A., April, C., Klotzle, B., Ota, T., Lingle, W. L., Gonzalez Bosquet, J., Shridhar, V., and Hartmann, L. C. (2009) Analysis of gene expression in stage I serous tumors identifies critical pathways altered in ovarian cancer *Gynecol Oncol* **114**, 3–11.
21. Cronin, M., Pho, M., Dutta, D., Stephans, J. C., Shak, S., Kiefer, M. C., Esteban, J. M., and Baker, J. B. (2004) Measurement of gene expression in archival paraffin-embedded tissues: development and performance of a 92-gene reverse transcriptase-polymerase chain reaction assay *Am J Pathol* **164**, 35–42.
22. Antonov, J., Goldstein, D. R., Oberli, A., Baltzer, A., Pirotta, M., Fleischmann, A., Altermatt, H. J., and Jaggi, R. (2005) Reliable gene expression measurements from degraded RNA by quantitative real-time PCR depend on short amplicons and a proper normalization *Lab Invest* **85**, 1040–50.

Chapter 7

MicroRNA Expression Analysis: Techniques Suitable for Studies of Intercellular and Extracellular MicroRNAs

Erica Hennessy and Lorraine O'Driscoll

Abstract

MicroRNAs, the class of small ribo-regulators, have been implicated in the regulation of a range of different biological processes, including development and differentiation, proliferation, and cell death. Only for a small fraction of identified microRNAs has a function been elucidated; therefore, a great deal of research remains to be performed to fully understand the role and implications of microRNAs.

This chapter discusses protocols for the isolation of microRNAs, reverse transcription, PCR, and large scale profiling using TaqMan low density miRNA arrays for analysis of microRNA expression levels.

Key words: MicroRNA, miRNA, TaqMan low density microRNA array, Megaplex RT primers

1. Introduction

MicroRNAs (miRNAs) are a family of endogenous small non-coding RNAs of 19–28 nucleotides in length that regulate gene expression at the post-transcriptional level by binding to complementary sites in the 3′-untranslated region (3′UTR) of target mRNAs. Based on the level of complementarity, miRNAs can direct target mRNAs for degradation or translational repression (1). MiRNAs are estimated to regulate at least one third of all human genes (2) and have been implicated in both normal physiological and pathological conditions, including differentiation and development, metabolism, proliferation, cell death, viral infection, and cancer (3), bringing miRNAs to the forefront of molecular biology interest.

A range of molecular biology tools have been developed in recent years for the identification, quantification, and functional

Lorraine O'Driscoll (ed.), *Gene Expression Profiling: Methods and Protocols*, Methods in Molecular Biology, vol. 784, DOI 10.1007/978-1-61779-289-2_7,

analysis of miRNAs both in vitro and in vivo. This chapter discusses the techniques used for the isolation and quantification of miRNA expression levels.

Many of the column-based total RNA isolation kits available result in loss of the small RNA fraction (i.e. RNAs of approximately less than 200 nucleotides in length) including miRNAs. Subsequently, many specialist miRNA isolation kits were developed, included the Applied Biosystem MirVana miRNA isolation kit which provides a relatively straightforward, quick procedure (approximately 30 min) for the isolation of total RNA including miRNAs, with the option of further enrichment specifically for small RNAs. For the procedures discussed in this chapter, however, enrichment for small RNAs is not required. Here, the TriReagent method for total RNA isolation (including miRNAs) is described due to the presence of this reagent in most molecular biology labs. This method is based on a guanidine thiocyanate phenol-chloroform extraction for the simultaneous extraction of total RNA, DNA, and protein (4). Total RNA is then precipitated using isopropanol; subsequent washes with 75% ethanol remove any contaminating traces of DNA or protein. The quality and quantity of isolated RNA can be assessed by measuring the ratio of absorbance at 260 and 280 nm. A ratio of 2 indicates good quality RNA, while a lower ratio indicates the presence of protein, phenol, or other contaminants.

The procedure for reverse transcription (RT) of miRNAs as described below involves the use of specific RT primers for the sequence-specific RT of individual miRNAs in a sample. TaqMan chemistries are then employed for real-time relative quantification of miRNA expression levels. TaqMan assays utilise a target-specific probe, containing a fluorescent reporter dye bound to the 5′ end, and a non-fluorescent quencher bound to the 3′ end. When a probe binds specifically to a target sequence, the fluorescent reporter dye is cleaved off due to the 5′–3′ exonuclease activity of *Taq DNA polymerase* enzyme. TaqMan probe-based chemistries provide an extra layer of specificity on SYBR green technology, which fluoresces when bound to double-stranded DNA. Exiqon offers an alternative miRNA PCR technology – miRCURY LNA (locked nucleic acids). miRCURY LNA miRNA PCR primers have incorporated a backbone modification of the sugar residues allowing for high affinity binding to target miRNAs. SYBR green technology is then used for detection of these double-stranded duplexes. miRCURY LNA and SYBR technologies are not discussed in detail in this chapter, for further information see http://www.exiqon.com/mirna-pcr.

Relative quantification of miRNA expression levels is calculated based on the ΔΔCt method (5). An endogenous control, which is expressed constitutively at the same level in all conditions under analysis, is used for normalisation of cycle threshold (C_T) values. A relevant calibrator sample, such as an untreated or time zero sample, is then used for calculation of relative expression levels in test samples.

Large scale miRNA profiling can be performed using TaqMan low density miRNA arrays (TLDAs). TLDAs are 384-well microfluidic cards, with each well representing a unique miRNA assay or endogenous control. A set of two cards are available for profiling expression levels of 754 miRNAs plus controls, based on miRBase v 14. Megaplex miRNA RT primer pools are used to reverse transcribe RNA for subsequent analysis using TLDAs. These primer pools contain two sets of RT primers, used for RT of the specific miRNAs on each of the TLDA cards. TLDA cards utilise 350–1,000 ng of RNA per card. Alternatively, as little as 1 ng RNA can be used if a pre-amplification step is performed. Exiqon also offer an alternative option for large scale miRNA profiling, their Universal cDNA Synthesis Kit incorporates a polyadenylation step allowing for RT of all miRNAs into cDNA. Ready-to-Use PCR panels are then used to profile expression of 742 miRNAs.

Currently (as of 11/07/2011), 1424 miRNAs have been identified in the human genome (miRBase v 17.0, http://www.mirbase.org/) (6–8). However, the function of only a small fraction of these miRNAs has been elucidated. The development of technologies for the analysis of miRNA expression and function, such as those described in this chapter will, undoubtedly, aid in our understanding of miRNAs and elucidating their role in the cell.

2. Materials

2.1. Preparation of Samples

1. Phosphate-buffered saline (PBS).

2.2. Isolation of RNA

1. TriReagent, store at 4°C.

   TriReagent is toxic and should be used under a fume cupboard.
2. Chloroform – Chloroform is harmful and should be used under a fume cupboard.
3. Isopropanol – Isopropanol is an irritant and is also highly flammable.
4. 75% Ethanol: Add 75 mL of 100% ethanol to 25 mL of water. Ethanol is highly flammable.
5. RNase-free water.

2.3. Generation of miRNA cDNA

1. RNase-free water.
2. TaqMan miRNA reverse transcription kit (containing 100 mM dNTPs, MultiScribe reverse transcriptase 50 U/μL, 10× RT buffer and RNase inhibitor 20 U/μL) (Applied Biosystems), store at −20°C.

3. TaqMan RT primers, specific for individual miRNAs, from TaqMan miRNA assays (Applied Biosystems), store at −20°C.
4. Thermocycler.

2.4. Real-Time PCR Amplification of miRNAs

1. TaqMan Universal PCR master mix, no amperase UNG (Applied Biosystems), store at 4°C.
2. Nuclease-free water.
3. TaqMan miRNA assays available for each specific miRNA to be analysed (Applied Biosystems), light sensitive, store at −20°C.
4. MicroAmp fast optical 96-well reaction plate with barcode 0.1 mL (Applied Biosystems).
5. MicroAmp optical adhesive film (Applied Biosystems).
6. 7900HT or 7500 fast real-time PCR instrument.

2.5. TaqMan Low Density miRNA Arrays

1. Megaplex RT primers (Applied Biosystems), store at −20°C.
2. TaqMan miRNA reverse transcription kit (containing 100 mM dNTPs, MultiScribe reverse transcriptase 50 U/μL, 10× RT buffer, and RNase inhibitor 20 U/μL) (Applied Biosystems), store at −20°C.
3. TaqMan universal PCR master mix, no amperase UNG (Applied Biosystems), store at 4°C.
4. TaqMan miRNA array (Applied Biosystems), light sensitive, store at 4°C.
5. 7900HT fast real-time PCR instrument.

3. Methods

3.1. Preparation of Samples

1. Cell pellets are prepared by centrifuging cell suspension at 1,000 × *g* for 5 min.
2. Discard supernatant. Resuspend pellet with 1 mL PBS and centrifuge at 1,000 × *g* for 5 min.
3. Repeat step 2.
4. Discard supernatant. Ensure all traces of PBS are removed. Store pellet at −80°C until required.

3.2. Isolation of RNA

1. Add 1 mL of TriReagent to 5–10 × 10^6 cells (see Note 1). Lyse cells by repeat pipetting and allow to stand at room temperature for 5 min.

2. Add 0.2 mL chloroform per mL of TriReagent. Shake vigorously for 15 s and allow to stand at room temperature for 15 min.
3. Centrifuge at 12,000×*g* for 15 min at 4°C.
4. Three phases should be visible in sample at this stage; the lower organic phase containing protein, an interphase containing DNA, and the upper aqueous phase containing RNA (see Note 2). Remove the upper aqueous phase to a fresh Eppendorf and add 0.5 mL of ice-cold isopropanol per mL of TriReagent used in step 1.
5. Mix sample by inverting and allow to stand at room temperature for 5–10 min (see Note 3).
6. Centrifuge at 12,000×*g* for 30 min at 4°C to pellet precipitated RNA.
7. Remove supernatant without disturbing the pellet.
8. Wash pellet with 1 mL of 75% ethanol. Vortex sample and centrifuge at 7,500×*g* for 5 min at 4°C.
9. Remove supernatant and repeat washing step 8.
10. Remove supernatant and allow RNA pellet to air-dry for 5–10 min (see Note 4).
11. Add an appropriate volume of RNase-free water, mix by repeat pipetting, allow to stand at room temperature for 10–15 min to allow dissolution of RNA pellet.
12. Assess RNA quality and quantity using NanoDrop (see Note 5).
13. Store at −80°C until required.

3.3. Generation of microRNA cDNA

1. Dilute total RNA isolated (as described in Subheading 3.2) to 2 ng/µL with RNase-free water. Add 5 µL of diluted RNA to Eppendorf tube. Use multiple tubes, if multiple miRNAs are to be reverse transcribed from the same sample (see Notes 6 and 7).
2. Prepare RT master mix as follows (volume for one sample):

100 mM dNTPs	0.15 µL
MultiScribe reverse transcriptase 50 U/µL	1.00 µL
10× Reverse transcription buffer	1.50 µL
RNase inhibitior 20 U/µL	0.19 µL
Nuclease-free water	4.16 µL
Total	7.00 µL

3. Mix RT master mix by flicking, centrifuge briefly to bring solution to the bottom of tube, and place on ice.
4. Add 7 μL of RT master mix to 5 μL of diluted RNA from step 1.
5. Mix gently by flicking and centrifuge to bring solution to bottom of tube (see Note 8).
6. Add 3 μL TaqMan RT primer to RT master mix/RNA tube (see Note 9).
7. Mix gently by flicking and centrifuge to bring solution to the bottom of tube.
8. Place sample in thermocycler and run the following cycle:
   (a) 16°C for 30 min.
   (b) 42°C for 30 min.
   (c) 85°C for 5 min.
9. Store miRNA cDNA at –20°C until required.

3.4. Real-Time PCR Amplification of miRNAs

1. Prepare PCR master mix as follows (volume for one well) (see Notes 10 and 11):

TaqMan 2× Universal PCR master mix	10.00 μL
Nuclease-free water	7.67 μL
Total	17.67 μL

2. Mix gently and centrifuge to bring solution to the bottom of the tube.
3. Add 1 μL of TaqMan miRNA assay to PCR master mix, mix gently and centrifuge.
4. Transfer 1.33 μL of RT product from Subheading 3.3 to PCR plate (see Notes 12 and 13).
5. Add 18.67 μL of PCR master mix to each sample of RT product, mix gently by repeat pipetting.
6. Seal the plate with an optical adhesive cover.
7. Run plate at the following cycle parameters on standard mode on 7500 or 7900HT real-time PCR instrument (If using a PCR master mix with Amperase UNG, incorporate a 50°C for 2 minutes step before step (a) to allow activation of Amperase UNG activity):
   (a) 95°C for 10 min.
   (b) 95°C for 15 s.
   (c) 60°C for 60 s.
   (d) Run 40 cycles of steps (b) and (c).
8. Analyse real-time PCR relative quantification data.

3.5. Large Scale miRNA Profiling Using TaqMan Low Density MicroRNA Arrays

1. Dilute RNA, 3 μL RNA to be used per reaction containing 350–1,000 ng (see Note 14).
2. Prepare Megaplex RT master mix as follows (volume for one array):

Megaplex RT primers (10×)	0.8 μL
dNTPs (100 mM)	0.2 μL
MultiScribe reverse transcriptase (50 U/μL)	1.5 μL
10× RT buffer	0.8 μL
$MgCl_2$ (25 mM)	0.9 μL
RNase inhibitor (20 U/μL)	0.1 μL
Nuclease-free water	0.2 μL
Total	4.5 μL

3. Mix gently and centrifuge to bring solution to the bottom of tube.
4. Add 3 μL of diluted RNA to Megaplex RT master mix.
5. Run sample on thermocycler at the following parameters:
   (a) 16°C for 2 min.
   (b) 42°C for 1 min.
   (c) 50°C for 1 s.
   (d) Repeat steps (a)–(c) for 40 cycles.
   (e) 85°C for 5 min.
   (f) Hold at 4°C.
6. Allow TaqMan miRNA array to warm to room temperature.
7. Prepare PCR reaction mix as follows (volume for one array):

TaqMan universal PCR master mix	450 μL
Megaplex RT product	6 μL
Nuclease-free water	444 μL
Total	900 μL

8. Mix gently and centrifuge to bring solution to the bottom of tube.
9. Dispense 100 μL of PCR reaction mix into each port of the TaqMan miRNA array (see Note 15).
10. Centrifuge TaqMan miRNA array cards at 331 × *g* for two consecutive 1 min spins.
11. Seal TaqMan miRNA array cards and load onto 7900HT Fast Real-Time PCR system (see Note 16).

12. Run TLDA cards at the following cycling parameters on standard mode:
    (a) 50°C for 2 min.
    (b) 94.5°C for 10 min.
    (c) 97°C for 30 s.
    (d) 59.7°C for 1 min.
    (e) Repeat steps (c) and (d) for 40 cycles.
13. Analyse real-time PCR relative quantification data.

4. Notes

1. If using serum specimens, add 750 μL TriReagent to 250 μL serum and continue as per Subheading 3.2 (see Note 3).
2. On removal of upper aqueous phase containing RNA, much care needs to be taken to avoid withdrawing contaminating DNAs from the interphase section. We recommend leaving a little of the upper aqueous phase containing RNA to minimise the possibility of carrying over DNA contamination.
3. If using samples with low levels of RNA, e.g. serum, on adding isopropanol, glycogen can also be added (to a final concentration of 120 μg/mL), and samples incubated overnight at −20°C to act as a carrier for precipitated RNA.
4. Do not allow pellet to air-dry completely, as this will decrease its solubility. Once pellet begins to change from a white to transparent colour then add RNase-free water to resuspend.
5. RNA purity is assessed by the ratio of absorbance at 260 and 280 nm. A ratio of approximately 2.0 indicates a good quality RNA sample. A lower ratio may indicate the presence of protein, phenol or other contaminants, which absorb at 280 nm, potentially carried over from the RNA isolation procedure.
6. When preparing RT master mix, multiply reagents by the appropriate number of miRNAs to be analysed and samples to be used. Always include 10–12% reagent excess when scaling up.
7. Minus RT controls i.e. without *MultiScribe reverse transcription* enzyme should be included, and used in the subsequent PCR experiments to ensure that the TaqMan miRNA assay does not pick up genomic DNA contaminants.
8. Do not exceed 350 × *g* or 5 min when centrifuging.
9. Before opening a tube of TaqMan RT primer, centrifuge to bring solution to the bottom of the tube and so to prevent loss of liquid trapped in the lid.

10. Prepare triplicate wells per sample per miRNA when performing real-time PCR.
11. Always perform endogenous control assay for each different sample used on the plate. "No template" controls should also be performed for each different miRNA assay being used on the assay plate.
12. Always add RT product to PCR plate first. In cases of confusion, it is obvious which wells contain 1.33 μL of RT product. If PCR master mix is added first, it can be very difficult to decipher to which wells RT product has been added to – as 18.67 μL and 20 μL look very similar.
13. When loading 96-well PCR plates, it may help to draw a template table, labelling which wells should contain which samples.
14. If RNA is limited or target miRNAs are expected to be expressed at low levels then very small amounts (1–350 ng) of RNA can be used for megaplex RT followed by a pre-amplification step before the real-time PCR.
15. Take care not to pierce the backing foil of the TLDA when dispensing PCR reaction mix into TaqMan miRNA arrays.
16. Take care that sealer and TaqMan miRNA array card are in the correct orientation before sealing the card, as sealing in the wrong orientation will rip the backing foil from the card and destroy the array.

References

1. Bartel, D. P. (2004) MicroRNAs: genomics, biogenesis, mechanism, and function. *Cell* ***116***, 281–97.
2. Zamore, P. D., and Haley, B. (2005) Ribo-gnome: the big world of small RNAs. *Science* ***309***, 1519–24.
3. Hennessy, E., and O'Driscoll, L. (2008) Molecular medicine of microRNAs: structure, function and implications for diabetes. *Expert Rev Mol Med* ***10***, e24.
4. Chomczynski, P., and Sacchi, N. (1987) Single-step method of RNA isolation by acid guanidinium thiocyanate-phenol-chloroform extraction. *Anal Biochem* ***162***, 156–9.
5. Livak, K. J., and Schmittgen, T. D. (2001) Analysis of relative gene expression data using real-time quantitative PCR and the 2(–Delta Delta C(T)) Method. *Methods* ***25***, 402–8.
6. Griffiths-Jones, S., Saini, H. K., van Dongen, S., and Enright, A. J. (2008) miRBase: tools for microRNA genomics. *Nucleic Acids Res* ***36***, D154–8.
7. Griffiths-Jones, S., Grocock, R. J., van Dongen, S., Bateman, A., and Enright, A. J. (2006) miRBase: microRNA sequences, targets and gene nomenclature. *Nucleic Acids Res* ***34***, D140–4.
8. Griffiths-Jones, S. (2004) The microRNA Registry. *Nucleic Acids Res* ***32***, D109–11.

Chapter 8

Western Blotting Analysis as a Tool to Study Receptor Tyrosine Kinases

Serena Germano and Lorraine O'Driscoll

Abstract

Receptor tyrosine kinases (RTKs) are involved in critical aspects of cell physiology ranging from cell survival, proliferation, growth, migration, and differentiation. A tight control of the extent and duration of signals elicited by activated RTKs is crucial for preventing over-stimulation, which can ultimately lead to unrestrained proliferative ability and neoplastic growth. Ligand-induced downregulation of RTKs has emerged as a key negative regulatory mechanism that can accomplish signaling attenuation, by removing activated receptors from the cell surface and committing them to degradation. The ability of RTKs to escape from ligand-induced downregulation has been reported as a recurrent mechanism of oncogenic deregulation in cancer.

Western blotting procedures have been extensively proven as straightforward assays to evaluate protein expression levels and have been widely applied to study RTKs downregulation.

Key words: Protein, Western blotting, Receptor tyrosine kinases

1. Introduction

Deregulated activation of receptor tyrosine kinases (RTKs) has been extensively documented in different types of human tumors (1) and frequently correlates with poor responsiveness to conventional therapies. The tight control of signals elicited by activated RTKs is critical for preventing over-stimulation and receptor downregulation has been shown to play a key role in this process (2, 3).

The product of the proto-oncogene RON is one such RTKs, identified as the receptor for macrophage stimulating protein (MSP) belonging to the subfamily of the Hepatocyte Growth Factor (HGF) receptor (4, 5). On engagement by its cognate ligand, Ron activates multiple intracellular signaling pathways

Lorraine O'Driscoll (ed.), *Gene Expression Profiling: Methods and Protocols*, Methods in Molecular Biology, vol. 784,
DOI 10.1007/978-1-61779-289-2_8,

including Ras/MAPK, PI-3K/Akt, JNK/SAPK, β-Catenin, and NF-kB (6) and mediating cellular proliferation, survival, migration, and differentiation. In physiological conditions, Ron activation is a transient event, whereas its deregulated activation is involved in cancer progression and metastasis in humans (6, 7) and in murine models (8). The RON gene is normally transcribed at relatively low levels in cells of epithelial origins, while overexpression and subsequent aberrant activation have been observed in epithelial cancers (9–11). Downregulation of this receptor, therefore, critically contributes to the maintenance of normal cell phenotype. Moreover, targeting Ron expression by forcing its downregulation has been proposed as suitable tool for therapy of cancers where altered Ron signaling is involved (12). All these studies involve an extensive use of western blotting technique to monitor Ron receptor expression levels.

Widely used in research and clinical laboratories, western blotting is an analytical method that involves the immobilization of proteins on membranes before detection with monoclonal or polyclonal antibodies. As it is based on a good resolution protein separation procedure such as sodium dodecylsulphate polyacrylamide denaturing gel (SDS–PAGE) and on high affinity antibody-antigen interactions, western blotting is a very sensitive and specific technique. Western blotting can provide qualitative information about the presence, the size, and post-translational modifications of proteins, but can also be used in a quantitative way to detect the relative abundance of a given protein in a sample. Since its original invention (13), this technique has improved and vast information exists on troubleshooting and improving the sensitivity, speed, and quantification. A typical western blotting procedure includes the following steps: (1) extraction and quantification of protein samples; (2) resolution of protein samples by electrophoresis on SDS–PAGE; (3) transfer of the separated polypeptides to a membrane support; (4) blocking nonspecific protein binding sites; (5) addition of antibodies; and (6) detection.

A detailed protocol to quantitatively study Ron ligand-dependent and ligand-independent downregulation by western blotting analysis and general methodological conditions optimized for the detection of RTKs expression are reported in this chapter.

2. Materials

2.1. Cell Culture, Extraction, and Quantification of Protein Samples

1. FG2 pancreatic carcinoma cells – RMPI-1640 medium supplemented with 10% fetal bovine serum (FBS) and 2 mM L-glutamine. NIH-3T3 fibroblast stably expressing Ron (3T3-Ron)-DMEM medium supplemented with 10% FBS and 2 mM L-glutamine.

2. Trypsin solution: 0.5 g/L porcine trypsin and 0.2 g/L EDTA·4Na in Hank's Balanced Salt Solution with phenol red.
3. Phosphate-buffered saline (PBS): sodium chloride 8 g/L, potassium chloride 0.2 g/L, di-sodium hydrogen phosphate 1.15 g/L, and potassium dihydrogen phosphate 0.2 g/L, pH 7.3 at 25°C.
4. MSP (R&D System, Minneapolis, MN), geldanamycin (GA) (Alexis, Montreal, Canada), and 17-allylamino-17-demethoxygeldanamycin (17-AAG) (Alexis).
5. Radioimmune precipitation buffer (RIPA): 50 mM Tris–HCl, pH 7.4, 150 mM NaCl, 0.5% sodium deoxycholate, 1% Triton X-100, and 0.1% SDS. Protease and phosphatase inhibitors (10 μg/mL aprotinin, 10 μg/mL leupeptin, 10 μg/mL pepstatin, 1 mM phenylmethylsulfonyl fluoride, 1 mM sodium orthovanadate, and 2 mM sodium fluoride) to be added immediately before use.
6. Micro-BCA protein assay kit (ThermoFisher Scientific Inc, Waltham, MA).
7. 6-Well cell culture clusters and polystyrene cell scrapers 18-mm blade.

2.2. SDS–Polyacrylamide Gel Electrophoresis

1. Laemmli sample buffer (2×): 4% SDS, 20% glycerol, 10% 2-mercaptoethanol, 0.004% bromphenol blue, and 0.125 M Tris–HCl, pH 6.8. Stored at –20°C.
2. Resolving buffer (4×): 1.5 M Tris–HCl, pH 8.8, 0.4% SDS. Stored at room temperature.
3. Stacking buffer (4×): 0.5 M Tris–HCl, pH 6.8, 0.4% SDS. Stored at room temperature.
4. Acrylamide/bis-acrylamide, 30% solution (Sigma). Stored at 4°C.
5. *N,N,N,N'*-Tetramethyl-ethylenediamine (TEMED) (Sigma). Stored at 4°C.
6. Ammonium persulfate: 10% solution in double-distilled water, freshly prepared.
7. Running buffer (10×): 250 mM Tris, 1.92 M glycine, and 1% (w/v) SDS. Stored at room temperature.
8. Prestained Molecular weight markers: PageRuler™ Prestained Protein Ladder (Fermentas, Burlington, Canada).
9. 1-D Electrophoresis system (Bio-Rad Laboratories Inc., Hercules, CA).

2.3. Western Blotting

1. Blotting buffer: 25 mM Tris–HCl pH 8.3, 192 mM glycine and 20% (v/v) methanol. Stored at 4°C. Methanol to be added immediately before use.

2. Immun-Blot PVDF membrane and extra thick blot paper (Bio-Rad Laboratories Inc.).
3. Tris-buffered saline (TBS, 10×): 100 mM Tris–HCl pH 7.5, 1.5 M NaCl. Stored at room temperature.
4. Blocking buffer: 5% (w/v) bovine serum albumin (BSA) solution in TBS. Single aliquots frozen at –20°C.
5. Washing buffer (TBS-T): 1× TBS solution supplemented with 1% Tween-20. Stored at room temperature.
6. Antibody dilution buffer: 1× TBS supplemented with 3% (w/v) BSA and 1% Tween-20. Single aliquots frozen at –20°C.
7. Primary antibodies: Polyclonal antibody against Ron C-terminal domain (C-20) (Santa Cruz Biotechnology Inc., Santa Cruz, CA) diluted 1:200 in antibody dilution buffer and monoclonal anti-α-tubulin (B-5-1-2) (Sigma) diluted 1:5,000 in antibody dilution buffer before use.
8. Secondary antibodies: Anti-rabbit and anti-mouse IgG HRP-linked (Cell Signaling Technologies Inc., Danvers, MA). To be diluted 1:1,000 in antibody dilution buffer before use.
9. Enhanced chemiluminescent (ECL) reagents: Immobilon Western Chemiluminescent HRP Substrate (Millipore, Billerica, MA).
10. Stripping buffer: 62.5 mM Tris–HCl, pH 6.8, 2% (w/v) SDS. Stored at room temperature. Warm to 70°C and add 100 mM β-mercaptoethanol before use.
11. Wet electroblotting system (Bio-Rad Laboratories Inc.).
12. Imaging system: Molecular Imager Versa-Doc MP system (Bio-Rad Laboratories Inc.).

3. Methods

3.1. Cell Lysis and Samples Preparation

The effectiveness of the western blot analysis is strictly dependent on the quality of protein lysates prepared for polyacrylamide gel electrophoresis (SDS–PAGE). There are a number of available cell lysis buffers (14) that can be used depending on whether an antigen is primarily extracellular, cytoplasmic, or membrane-associated. The use of the appropriate detergent for protein extraction is, therefore a key step to obtain reliable results. RIPA lysis buffer is one of the most used buffers to lyse cultured mammalian cells from both plated cells and cells pelleted from suspension cultures. It contains two ionic detergents, SDS and sodium deoxycholate, as well as a non-ionic detergent Triton X-100, critical for denaturing proteins, breaking many protein–protein interactions and enabling the extraction of cytoplasmic, membrane, and nuclear proteins. To obtain reliable and reproducible results, quantification of protein content in cell extracts with a good precision is mandatory to proceed with quantitative studies. Indeed, even if it is possible to normalize for small differences in total protein by the use of housekeeping genes (e.g., α-tubulin), significant discrepancies in protein loading results in defective results of the western blotting analysis. It is thus of great importance to select an appropriate protein quantification assays. The procedure based on the use of bicinchoninic acid (BCA) has proved to be particularly suitable to use with cell lysates, given its compatibility with most ionic and non-ionic detergents.

1. For ligand-dependent downregulation studies, pancreatic carcinoma FG2 cells are seeded at sub-confluent density on 6-well plates (one well is required for each experimental data point) in normal growth medium and incubated for 24 h. The cultures are then rinsed twice with PBS and incubated for a further 24 h in medium without serum (see Note 1). Treatments with vehicle (−) or different concentration of MSP are performed for the indicated times (see Fig. 1a, b) and the cultures are then lysed.
2. For drug-induced downregulation studies, NIH-3T3 fibroblasts (3T3-Ron) and FG2 cells, expressing recombinant and endogenous Ron, respectively, are seeded at sub-confluent density on 6-well plates (one well is required for each experimental data point) in normal growth medium and incubated for 24 h. The cultures are then treated with vehicle (−), GA or 17-AAG (see Fig. 2) for the indicated times and then lysed.
3. For cell lysis, the cultures are carefully rinsed twice with PBS and ice-cold RIPA buffer, containing protease and phosphatase inhibitors (see Note 2), is added directly on cultured cells (150 μL RIPA buffer are required for each well). Plates are incubated on ice for 5 min with occasional swirl.
4. Bottom of wells is gently scraped with clean cell scrapers and cell lysates are collected in microcentrifuge tubes.
5. Lysates are then cleared by centrifugation at 13,000 × *g* for 15 min at 4°C and supernatants are transferred to fresh tubes. This separates total proteins (supernatant) from the cellular debris (pellet).
6. Five microliter aliquots are used for protein quantification. The lysates are stored at −80°C, if not used immediately (see Note 3).
7. Total protein content of the samples is determined using the Micro BCA Protein assay kit (see Note 4). Before starting, preheat a hot-block or water bath to 60°C.

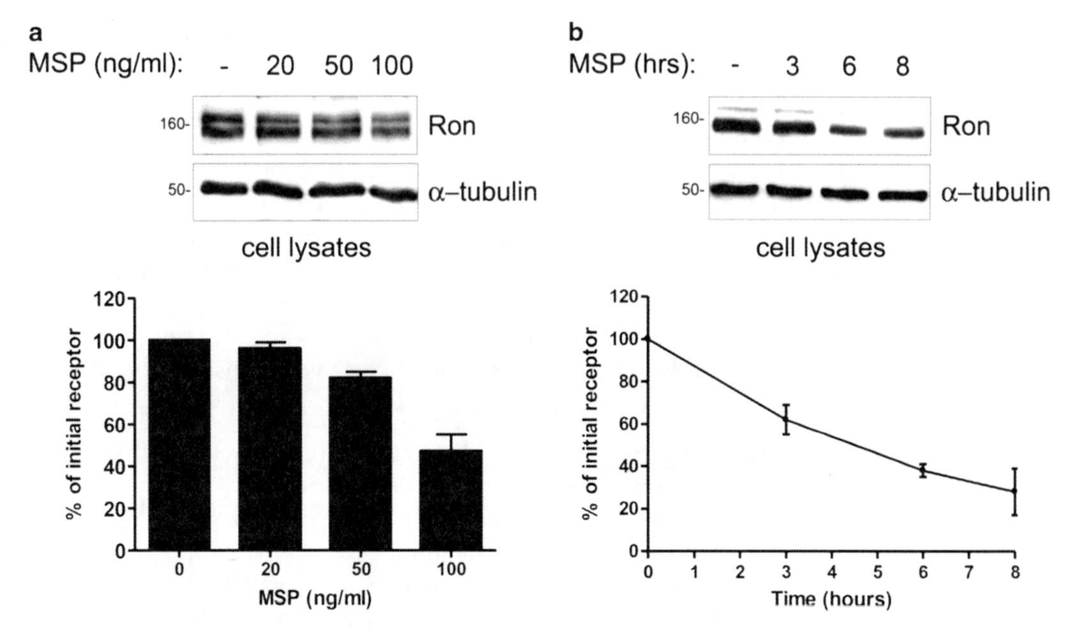

Fig. 1. Ligand-induced Ron downregulation. (**a**) FG2 cells endogenously expressing Ron are treated with vehicle (–) or with the indicated amount of human recombinant MSP for 8 h. Cell lysates are analyzed by western blotting with a polyclonal antibody against Ron C-terminal domain, recognizing both the 170-kDa precursor and the 150-kDa β-chain of the receptor. α-Tubulin blotting is used as loading control. In the *lower panel*, densitometry of Ron immunoblotting is shown and values, normalized to the relative α-tubulin control bands, are plotted as percentages of vehicle-treated cells. Each data point represents the mean ± SE of three independent experiments. (**b**) Time course of MSP-induced Ron downregulation. FG2 cells are treated with 100 ng/mL MSP for the indicated times or left untreated (0 h). Cell lysates are analyzed by western blotting as described in panel (**a**). In the *lower panel*, densitometry of Ron is shown and values, normalized to the relative α-tubulin control bands, are plotted as percentages of unstimulated cells. Each data point represents the mean ± SE of three independent experiments.

8. Calculate the total volume (TV) of BCA Working Solution for the unknowns samples, six standards and 1 mL excess [TV = (# unknowns + 6 standards + 1) × 0.5 mL]. Use 25 parts of reagent A, 24 parts of reagent B, and 1 part of reagent C, using the following formula: $V_A = TV/50 \times 25$; $V_B = TV/50 \times 25$; $V_C = TV/50$. Mix the reagents thoroughly.
9. Prepare the test samples: dispense 495 μL of water in 1.5-mL test tubes and then add the 5 μL cell extract aliquots from step 6 to each tube.
10. Prepare the BSA standards at the following final concentrations: 0 μg/mL (blank), 1, 2.5, 5, 10, and 20 μg/mL. Dispense the following amounts of water in 1.5-mL test tubes: 500, 499, 497.5, 495, 490, and 480 μL. Then add to the tubes 0, 1, 2.5, 5, 10, and 20 μL of 1 μg/μL BSA solution, respectively.
11. Add 500 μL of the working solution to each test tube, mix thoroughly and incubate the tubes in the heat-block or water bath at 60°C, with the caps closed.

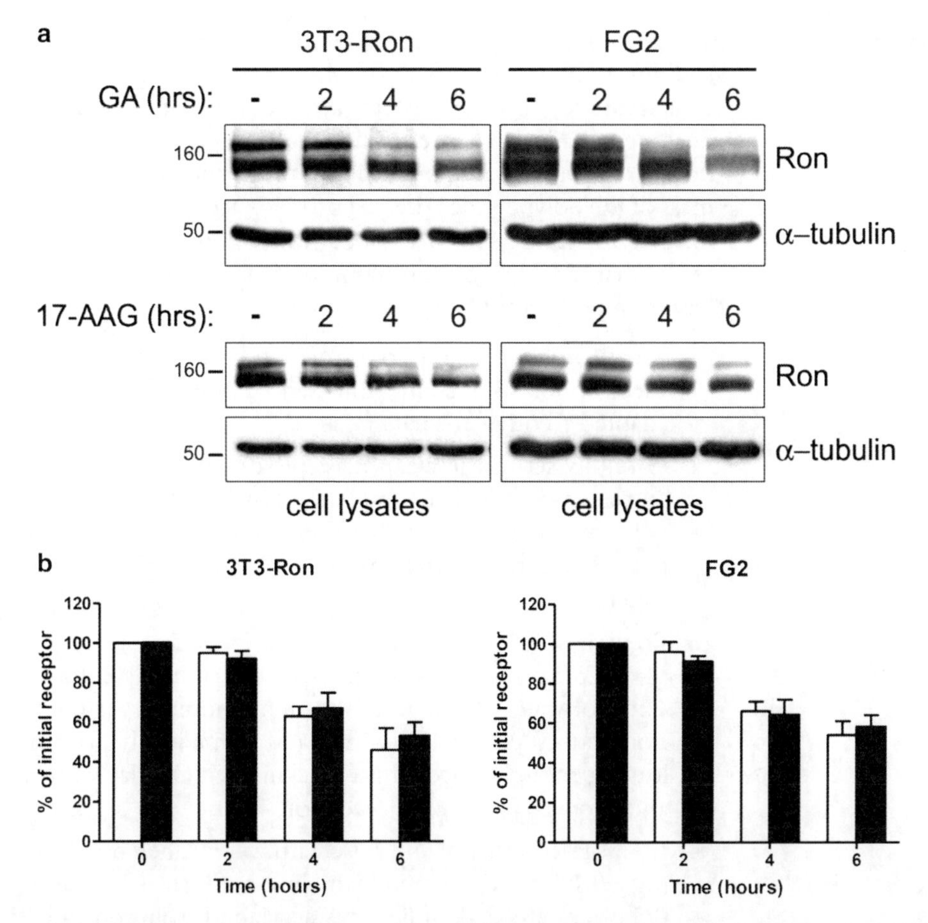

Fig. 2. (**a**) Geldanamycins induce degradation of both precursor and mature cell surface Ron. 3T3-Ron and FG2 cells are treated with vehicle (–) or 1 μm GA or 1 μm 17-AAG for the indicated times. Cell lysates are analyzed by western blotting with a polyclonal antibody against Ron C-terminal domain. α-Tubulin blotting is used as loading control. (**b**) Densitometry of Ron immunoblotting is shown and values, normalized to the relative α-tubulin control bands, are plotted as percentages of vehicle-treated cells. *White columns*: GA; *black columns*: 17-AAG. Each data point represents the mean ± SE of three independent experiments.

12. Cool the tubes at room temperature for 5 min, then transfer the content in plastic cuvettes and immediately read absorbance at 562 nm, with the spectrophotometer zeroed using a cuvette containing the 0 μg/mL standard.
13. Prepare a standard curve by plotting the Abs for each standard against its concentration. Use the standard curve to calculate the protein concentration of each unknown sample.
14. Preheat the hot-block to 95°C. Transfer 25–30 μg of total protein samples from step 5 to new tubes and add an equal amount of Laemmli sample buffer (2×) (see Note 5). Close the lids and incubate the samples at 95°C for 5 min to denaturate the proteins. Cool down to room temperature before loading the gel (see Note 6).

3.2. SDS–PAGE

The SDS–PAGE procedure utilizes the strongly anionic detergent sodium dodecyl sulfate with or without a reducing agent and heat to denaturate proteins. Proteins bind to SDS within and as a result become negatively charged. The amount of SDS that binds to the denatured proteins is approximately proportional to the molecular mass of the polypeptides (i.e., all proteins acquire the same charge/mass ratio) facilitating directional migration of the highly negatively charged SDS–protein complexes in the polyacrylamide gel based on polypeptide size.

1. The protocol here reported refers to the use of mini gels 1-D Electrophoresis system (Bio-Rad Laboratories), but can be adapted to other formats.
2. Carefully scrub the clean glass plates with 95% (v/v) ethanol and assembly the front and back glasses in the apposite clamps. It is advisable to check that there is no leakage from the bottom by pouring water inside the plates. Then pour off the water.
3. Prepare a 7.5% gel solution by mixing 2 mL of resolving buffer (4×), 2 mL of 30% acrylamide/bis-acrylamide solution, and 4 mL of water. Then add 100 μL ammonium persulfate solution and 40 μL TEMED, mix and immediately pour the gel, leaving enough space for the stacking gel. Overlay with ethanol and allow to polymerize (see Note 7).
4. Pour off the ethanol and rinse with water. Then pour the stacking gel solution prepared by mixing 1 mL stacking buffer (4×), 0.5 mL of 30% acrylamide/bis-acrylamide solution and 2.5 mL of water. Immediately insert the combs and allow to polymerize (see Note 7).
5. Remove the glasses from the holder, then assemble the gasket with the electrodes. Remove the comb, fill the gasket and the outer chamber with running buffer (1×), and then wash the wells with a syringe fitted with a thin gauge needle.
6. Load the samples and the molecular markers (see Note 8) in the wells.
7. Assembly the unit and connect to the power supply. Use constant voltage up to 130 V through the stacking gel, then up to 150 V in the running gel. Turn off the power supply immediately after the bromophenol blue dye has run off the gel and disconnect the power supply.

3.3. Western Blotting

This section describes the electrotransfer of proteins separated from SDS–PAGE gels via a step-by-step procedure.

1. The procedure here described assumes the use of a wet transfer method (see Note 9) employing a Bio-Rad Trans-Blot system.

2. Wet a sheet of PVDF paper, cut slightly larger than the gel size, in a tray containing methanol for 1 min to activate the membrane. Then transfer it in another tray filled with blotting buffer. Wet two sheets of extra thick paper and two sponges in blotting buffer.
3. Disassemble the gel unit, cut and remove the stacking gel with a blade, and then transfer the resolving gel in a tray containing blotting buffer. If desired, a corner of the gel can be cut to allow the tracking of the gel orientation.
4. Assembly the transfer cassette by lying a sheet of paper onto a sponge and the PVDF membrane on the top. The gel is then carefully laid in top of the membrane. Then another sheet of thick paper and a sponge are positioned on top of the gel and the transfer cassette is locked. Make sure that no air bubbles remain between the gel and the membrane.
5. Insert the cassette in the transfer tank, carefully checking the orientation, the membrane must be oriented towards the anode, while the gel towards the cathode (see Note 10). Put an iced freezer pack in the tank, add enough blotting buffer to cover the cassette and activate a magnetic stir-bar in the tank to avoid heating of the buffer.
6. Close the lid, connect the unit to the power supply and begin transfer, with a constant current of 200 mA for 1.5 h.
7. Disconnect the power supply; then disassemble the transfer cassette. Remove the sponge, the paper, and the gel. If using pre-stained markers, check that their corresponding bands are clearly visible on the membrane.
8. Add to the membrane 10 mL of Ponceau S staining solution and incubate the membrane for 2 min. The Ponceau S solution can be reused several times.
9. Wash the membrane briefly with TBS to remove excess staining. Continue washing the membrane until the staining is gone. If stains persist, wash the membrane TBS containing 0.02% NaAzide for 1–3 min and then rinse once with TBS.
10. The membrane is then incubated in 15 mL blocking buffer for 1 h at room temperature on a rocker with gentle shaking (i.e., 20 rpm).
11. After blocking, the membrane is rinsed twice with TBS and then incubated for 3 h with primary antibody solution at room temperature with gentle shaking. Alternatively, the incubation can be performed overnight at 4°C to enhance the signal.
12. The primary antibody is then removed (see Note 11) and the membrane is washed three times for 10 min each with 15 mL of TBS-T with vigorous shaking (i.e., 40 rpm).

13. Freshly prepared secondary antibody solution is then added to the membrane for 1 h at room temperature with gentle shaking.
14. The secondary antibody solution is discarded and three washes for 10 min each with TBS-T are performed with vigorous shaking.
15. The ECL reagents are mixed together in the ratio 1:1 immediately before use and evenly added to the blot for 3 min.
16. The excess of ECL is removed and the membrane is put in a tray. Proceed with image acquisition as detailed in the next section.
17. Once a satisfactory signal has been obtained, wash the membrane and then proceed with the stripping procedure to clear the membrane before reprobing for a housekeeping gene (see Note 12).
18. For the stripping procedure, warm 30 mL stripping buffer at 70°C and then add. Incubate the membrane in this solution for 30 min and then perform extensive washes with TBS. Repeat the blocking step again before reprobing with the primary antibody solution.

3.4. Quantification by Image Densitometry

Digital imaging systems are preferably used in protein expression analysis. Indeed, they display a higher dynamic range when compared to autoradiography films (15). Moreover, digital imaging allows rapid data digitization and monitoring of image saturation immediately after the acquisition. Systems equipped with flat field correction give best results, due to a lower coefficient of variation along the image.

1. A suitable exposure for reliable data quantification must not present saturated signals. This could be easily checked when using digital imaging system is a quantitative imaging system (see Note 13). If using autoradiography sheets, acquire the image by scanning (a scanner that has a transparency mode gives better results).
2. Once a satisfactory image has been obtained, proceed with receptor bands quantification using an Image Analysis Software.
3. Subtract background signals from all bands (see Note 14).
4. Repeat steps 3 and 4 quantifying the housekeeping gene bands (i.e., α-tubulin).
5. Calculate, using background-subtracted data, normalized values by dividing receptor bands densities by their respective housekeeping gene bands densities (i.e., Ron/α-tubulin).
6. Set normalized values obtained from the unstimulated cells as the calibrator.

7. Calculate the fraction of residual receptor expression after stimulation: divide all normalized values by the calibrator value (i.e., MSP-stimulated cells/unstimulated cells).
8. Alternatively, the percentage of receptor reduction can be easily calculated by the following formula: 100 × (calibrator – sample)/calibrator (i.e., 100 × (unstimulated cells – MSP-stimulated cells)/unstimulated cells).

4. Notes

1. Serum starvation is required when studying ligand-induced receptor downregulation, as growth factor contained in serum can bias the results obtained.
2. Proteases can be released during lysis, as well as phosphatases, and act on the target proteins. Therefore, to prevent proteolysis, protease and phosphatase inhibitors should be included in the lysis buffer and added immediately before applying it to the cultures. Inhibitors are available separately or can be purchased as cocktails in tablet form.
3. To avoid protease degradation of proteins in the samples, cell lysis should be performed quickly on ice and cell lysates should be immediately stored at –80°C and further freeze and thaw should be avoided.
4. The linear working range of this procedure is 0.5–20 μg/mL. Several kits based on BCA, with different linearity and volume formats, are available on the market.
5. Note that up to 70 μL can be accommodated in each well of the gel, when using 1.5-mm spacers and 10-well combs, so consider lowering the total amount of proteins to be loaded in case of low concentration samples (but avoid quantity lower than 10–15 μg).
6. Boiled samples can be stored at –80°C or –20°C. Boil again at 95°C for 1 min when thawing, to facilitate SDS resolubilization.
7. Polymerization is complete when the extra gel solution that has not been poured and has been kept in the tube is completely polymerized.
8. Molecular markers based on recombinant proteins have the advantage of lot-to-lot consistency. Prestained molecular weight marker standards are available from several companies. Markers containing protein conjugated to different dyes (multicolor markers) allow an easier identification of the bands.
9. Transfer tank blot systems (also known as wet transfer) offer a more efficient transfer of high molecular weight proteins (16),

such as RTKs. Moreover, with the semi-dry system, some molecules will not transfer quantitatively.

10. The Trans-Blot system from Bio-Rad is endowed with a black and white color code to check the correct orientation of the cassette.
11. Primary antibody solutions can be used several times. It is possible to store them up to 1 month at 4°C upon addition of 0.02% of NaAzide. Nevertheless, it should be taken into consideration a progressive loss of signal due to antibody depletion after each incubation step.
12. When using a housekeeping gene with a molecular weight significantly different from the gene of interest (as in the case of α-tubulin and Ron), it is possible to proceed with the second staining step without stripping the membrane.
13. When using the Quantity One Software with a GelDoc or VersaDoc System, verify that the "highlight saturated pixel" option is ticked: saturated pixels will turn red.
14. Image analysis software offers many option to perform background calculation and subtraction. An easy way to do it, is to select an empty area of the membrane of the same size of an average band, and use as background reference.
15. This protocol can be adapted for many other protein quantification issues. For instance, it can be used to monitor protein overexpression/silencing in gain-of- or loss-of-function experiments.

Acknowledgments

Preparation of this chapter was supported by the Health Research Board of Ireland (Grant RP/2006/77); Trinity College Dublin's Start-Up Funds for New Academics 2008/2009 and Science Foundation Ireland's funding of Molecular Therapeutics for Cancer, Ireland (08/SRC/B1410).

References

1. Blume-Jensen, P., Hunter, T. (2001) Oncogenic kinase signalling. *Nature.* **411**, 355–365.
2. Dikic, I., Giordano, S. (2003) Negative receptor signalling. *Curr Opin Cell Biol.* **15**, 128–135.
3. Peschard, P., Park, M. (2003) Escape from Cbl-mediated downregulation: a recurrent theme for oncogenic deregulation of receptor tyrosine kinases. *Cancer Cell.* **3**, 519–523.
4. Gaudino, G., Follenzi, A., Naldini, L., Collesi, C., Santoro, M., Gallo, K.A., Godowski, P.J., Comoglio, P.M. (1994) RON is a heterodimeric tyrosine kinase receptor activated by the HGF homologue MSP. *EMBO J.* **13**, 3524–3532.
5. Wang, M.H., Ronsin, C., Gesnel, M.C., Coupey, L., Skeel, A., Leonard, E.J., Breathnach, R. (1994) Identification of the ron gene product as the

receptor for the human macrophage stimulating protein. *Science.* **266**, 117–119.

6. Germano, S., Gaudino, G. (2008) Molecular targets in cancer therapy: the Ron approach. *Oncology Rev.* **1**, 215–224.
7. Wang, M.H., Wang, D., Chen, Y.Q. (2003) Oncogenic and invasive potentials of human macrophage-stimulating protein receptor, the RON receptor tyrosine kinase. *Carcinogenesis.* **24**, 1291–1300.
8. Peace, B.E., Toney-Earley, K., Collins, M.H., Waltz, S.E. (2005) Ron receptor signaling augments mammary tumor formation and metastasis in a murine model of breast cancer. *Cancer Res.* **65**, 1285–1293.
9. Willett, C.G., Wang, M.H., Emanuel, R.L., Graham, S.A., Smith, D.I., Shridhar, V., Sugarbaker, D.J., Sunday, M.E. (1998) Macrophage-stimulating protein and its receptor in non-small-cell lung tumors: induction of receptor tyrosine phosphorylation and cell migration. *Am J Respir Cell Mol Biol.* **18**, 489–496.
10. Zhou, Y.Q., He, C., Chen, Y.Q., Wang, D., Wang, M.H. (2003) Altered expression of the RON receptor tyrosine kinase in primary human colorectal adenocarcinomas: generation of different splicing RON variants and their oncogenic potential. *Oncogene.* **22**, 186–97.
11. Cheng, H.L., Liu, H.S., Lin, Y.J., Chen, H.H., Hsu, P.Y., Chang, T.Y., Ho, C.L., Tzai, T.S., Chow, N.H. (2005) Co-expression of RON and MET is a prognostic indicator for patients with transitional-cell carcinoma of the bladder. *Br J Cancer.* **92**, 1906–1914.
12. Germano, S., Barberis, D., Santoro, M.M., Penengo, L., Citri, A., Yarden, Y., Gaudino, G. (2006) Geldanamycins trigger a novel Ron degradative pathway, hampering oncogenic signaling. *J Biol Chem.* **281**, 21710–21719.
13. Burnette, W.N. (1981) "Western blotting": electrophoretic transfer of proteins from sodium dodecyl sulfate–polyacrylamide gels to unmodified nitrocellulose and radiographic detection with antibody and radioiodinated protein. *A. Anal Biochem.* **112**, 195–203.
14. Harlow, E. and Lane, D. (1999) Using antibodies: A laboratory manual, Cold Spring Harbor Laboratory Press.
15. Martin, C.S. and Bronstein, I. (1994) Imaging of chemiluminescent signals with cooled CCD camera systems. *J Biolumin Chemilumin.* **9**, 145–153.
16. MacPhee, D.J. (2009) Methodological considerations for improving Western blot analysis. *J Pharmacol Toxicol Methods.* **61(2)**, 171–177.

Chapter 9

2D Gel Electrophoresis and Mass Spectrometry Identification and Analysis of Proteins

Paula Meleady

Abstract

Analysis of the protein expression patterns in clinical samples and cells by proteomic technologies offers opportunities to discover potentially new biomarkers for early detection and diagnosis of disease. One of the most widely used techniques to study the proteome of a biological system is two-dimensional polyacrylamide gel electrophoresis (2D-PAGE). In particular, a modified version of 2D-PAGE, two-dimensional difference gel electrophoresis (2D-DIGE), which uses differential labelling of protein samples with up to three fluorescent tags, offers greater sensitivity and reproducibility over conventional 2D-PAGE. In this chapter, we will introduce methods for the analysis of whole cell lysates from human cancer cell lines using 2D-DIGE and identification of differentially expressed proteins using liquid chromatography mass spectrometry, i.e. LC–MS/MS.

Key words: Two-dimensional gel electrophoresis, Two-dimensional difference gel electrophoresis, Mass spectrometry, Protein expression profiling, Proteomics

1. Introduction

Monitoring the protein expression patterns in clinical samples and cells by proteomics technologies offers opportunities to discover potentially new biomarkers for early detection and diagnosis of diseases (1). Two-dimensional polyacrylamide gel electrophoresis (2D-PAGE) is one of the most widely used techniques to study the proteome of a cell or tissue (reviewed (2–4)). 2D-PAGE remains challenging, mainly because of its low sensitivity and reproducibility. There are also many limitations inherent to the technique including difficulties in separating low abundant proteins, hydrophobic proteins (e.g. membrane), high/low molecular weight proteins, and also proteins with extremes of pI. Modified 2D electrophoresis by fluorescent tagging to proteins, two-dimensional

Lorraine O'Driscoll (ed.), *Gene Expression Profiling: Methods and Protocols*, Methods in Molecular Biology, vol. 784,
DOI 10.1007/978-1-61779-289-2_9, © Springer Science+Business Media, LLC 2011

difference gel electrophoresis (2D-DIGE), offers increased throughput, ease of use, reproducibility, and accurate quantification of protein expression differences (5). This system enables the separation of two fluorescently labelled protein samples (Cy3 and Cy5) on the same gel. Gel-to-gel variability can be reduced using a Cy2-labelled internal standard on each gel of the experiment (6). Differential analysis software then allows the determination of differentially expressed protein targets from the gel images with statistical significance. Proteins can be enzymatically digested with trypsin to yield peptide fragments that can be readily analysed using Matrix Assisted Laser Desorption Ionization Time-of-Flight Mass Spectrometry (MALDI-ToF MS) to generate peptide mass fingerprints (PMF) for protein identification. Additionally, other mass spectrometry techniques such as electrospray ionisation (ESI-MS/MS) are capable of providing amino acid sequence information on peptide fragments of the parent protein (7).

To date, 2D-PAGE and mass spectrometry have been successfully applied to the differential proteomic analysis of various types of biological samples including tissue (8), serum (9), saliva (10), and cell line models (11, 12) to better understand the molecular basis of the pathogenesis of disease. In this chapter, we will introduce methods for the differential analysis of protein expression in human cancer cell lines using 2D-DIGE. We will also introduce methods for the mass spectrometry identification of differentially regulated proteins, in particular using LC–MS/MS.

2. Materials

2.1. Sample Preparation for 2D-PAGE

1. Phosphate-buffered saline (PBS).
2. Sucrose buffer: 10 mM sucrose, 100 mM Tris–HCl, pH 8.0. Prepare fresh on the day of use (see Note 1).
3. 2D-DIGE lysis buffer: 7 M urea, 2 M thiourea, 4% CHAPS, 30 mM Tris–HCl, 5 mM magnesium acetate, pH 8.5. Aliquot into 1 mL amounts and store at –20°C for up to 3 months (see Note 2).
4. Protein assay. Bradford assay kit (Bio-Rad).
5. 21-guage needles.
6. pH paper.
7. Refrigerated microcentrifuge.

2.2. 2D Difference Gel Electrophoresis

2.2.1. Labelling of Proteins with Fluorescent Cy Dyes

1. Cy2, Cy3, and Cy5 minimal labelling kits (GE Healthcare).
2. Dimethylformamide (DMF), water-free (Sigma) (see Note 3).
3. Lysine solution (10 mM). Prepare fresh on the day of use.
4. 2× sample buffer: 7 M urea, 2 M thiourea, 4% CHAPS, 40 mM Tris–HCl. Store in 1 mL aliquots at –20°C for up to 6 months. Add 2% DTT and 2% IPG buffer just before use.
5. IPG buffer, various pH ranges to suit the IPG strips that are being used (GE Healthcare).

2.2.2. First Dimension IEF

1. Immobiline pH gradient (IPG) strips (GE Healthcare). Various lengths and pH ranges available. Also available from other suppliers (e.g. Bio-Rad, Sigma).
2. Immobiline Dry Strip Reswelling tray (GE Healthcare).
3. Rehydration buffer: 7 M urea, 2 M thiourea, 4% (w/v) CHAPS, 40 mM Tris–HCl. Add a few grains of bromophenol blue to colour the solution. Store in 1 mL aliquots at –20°C for up to 6 months. Before use, add 3 mg of DTT and 2% (v/v) IPG buffer to 1 mL of rehydration buffer.
4. IPG Cover Fluid (GE Healthcare).
5. Sample cups (GE Healthcare).
6. Paper wicks (GE Healthcare).
7. IPGphor IEF unit including cup loading manifold (GE Healthcare).

2.2.3. Second Dimension SDS–PAGE

1. 12.5% acrylamide gel solution. Prepare 900 mL for casting 14 gels: 375 mL of 30% acrylamide/bisacrylamide solution (Sigma), 225 mL of 1.5 M Tris–HCl pH 8.8 (Bio-Rad), and 9 mL of 10% SDS solution (Sigma). Just before use add 9 mL of freshly prepared 10% ammonium persulphate solution (APS) and 125 μL of neat TEMED. Add the TEMED in a fume hood.
2. Displacing solution: Prepare 100 mL of 375 mM Tris–Cl pH 8.8, 50% glycerol. Add a few grains of bromophenol blue to colour the solution.
3. Water-saturated butanol.
4. SDS equilibration stock solution: 6 M urea, 30% glycerol, 50 mM Tris–Cl pH 8.8, 2% SDS, and a few grains of bromophenol blue to colour the solution. Aliquot into 30 mL volumes and freeze at –20°C for up to 6 months.
5. SDS equilibration buffer A: SDS equilibration stock solution containing 1% DTT. Dissolve the DTT just before use and do not store.
6. SDS equilibration buffer B: SDS equilibration stock solution containing 2.5% Iodoacetamide. Dissolve the iodoacetamide just before use and do not store.

7. Agarose overlay solution: 0.5% agarose in 2× SDS running buffer. Add a few grains of bromophenol blue to the melted agarose to colour the solution.
8. 10× SDS running buffer: prepare 10 L of 1.92 M glycine, 250 mM Tris–HCl, and 2% SDS and store at room temperature for up to 3 months. Do not adjust the pH of this solution, which should be approximately 8.3. Dilute the 10× buffer to obtain 1× and 2× SDS running buffer solutions.
9. Low fluorescent glass cassettes (GE Healthcare).
10. Ettan Dalt 12 electrophoresis system (GE Healthcare).
11. Fluorescent scanner/imager capable of imaging fluorescent gels, e.g. Typhoon 9400 Variable Mode Imager (GE Healthcare).
12. Software for cropping 2D gels, e.g. Image Quant (GE Healthcare).
13. 2D Differential Analysis software; various vendors including Decyder (GE Healthcare) and Progenesis SameSpots (NonLinear Dynamics).

2.3. Sample Preparation for MS Analysis from 2D-PAGE Gels (In-gel Digestion)

1. Gel stains: Coomassie Brilliant Blue (CBB) R250, silver stain, fluorescent stains such as Sypro Ruby, Deep Purple (see Note 4).
2. Solution containing 50% methanol and ammonium bicarbonate (40 mM). Prepare fresh on the day of use.
3. 30 mM potassium ferricyanide solution. Prepare fresh on the day of use.
4. 100 mM sodium thiosulphate solution. Prepare fresh on the day of use.
5. 100 mM ammonium bicarbonate solution. Prepare fresh on the day of use.
6. Solution containing 10 mM ammonium bicarbonate in 10% (v/v) acetonitrile. Prepare fresh on the day of use.
7. Solution containing 10 mM DTT in 100 mM ammonium bicarbonate. Make shortly before use.
8. Solution containing 55 mM iodoacetamide in 100 mM ammonium bicarbonate. Make shortly before use.
9. Trypsin solution: prepare 12.5 ng/μL trypsin in 10 mM ammonium bicarbonate containing 10% (v/v) acetonitrile. Resuspend the Trypsin Gold (sequencing grade, Promega) at 1 μg/mL in 50 mM acetic acid (100 μg vial in 0.1 mL of 50 mM acetic acid buffer). Remove the amount of this 80× trypsin stock needed and refreeze the unused portion in 10 μL aliquots and store at –20°C. Make the 1× trypsin shortly before use by diluting in 10 mM NH_4HCO_3 containing 10% (v/v) acetonitrile to a concentration of 12.5 ng/μL (1:80 (v/v)).
10. 20 mM ammonium bicarbonate solution. Prepare fresh on the day of use.

11. Solution containing 50% acetonitrile and 0.1% trifluoroacetic acid. Use a fume hood to prepare this solution.
12. 0.1% Triflouroacetic acid (TFA). Use a fume hood to prepare this solution.
13. Sonicating water bath.
14. Vacuum centrifuge.

2.4. LC–MS/MS for Protein Identification from 2D-PAGE Gels

1. Solvent A: 2% acetonitrile in LC–MS grade water containing 0.1% formic acid (prepare 1 L). Use a fume hood to prepare this solution.
2. Solvent B: 2% LC–MS grade water in acetonitrile containing 0.1% formic acid (prepare 1 L). Use a fume hood to prepare this solution.
3. Sample loading solution (i.e. trap column mobile phase: 0.1% TFA). Add 1 mL of TFA to 1 L of water. Use a fume hood to prepare this solution.
4. Nano LC System: Ultimate 3000 (Dionex/Thermo Fisher Scientific).
5. Mass Spectrometer: LTQ Orbitrap XL (Thermo Fisher Scientific).
6. Column: PepMap C18 capillary column (300 μm × 15 cm, 3-μm particles) (Dionex/Thermo Fisher Scientific).
7. Trap column: PepMap C18 trap cartridge (300 μm × 5 mm) (Dionex/Thermo Fisher Scientific).
8. Column oven.

3. Methods

3.1. Sample Preparation for 2D-PAGE

The sample preparation method described here is based on the analysis of whole cell lysates from cultured human cancer cell lines. Many sample preparation techniques exist for the analysis of in vitro and in vivo biological material, including fresh or frozen tissue, biological fluids (serum, saliva, etc.), and cultured mammalian cell lines, and are beyond the scope of this chapter.

1. Seed human cancer cells (e.g. MCF7, A549) into T-75cm^2 flasks at 5×10^5 cells per flask. Ensure that you set up enough biological replicate flasks in order to obtain statistically significant results.
2. Harvest exponentially growing cells and resuspend cell pellets in cold PBS (at 4°C). Centrifuge at $250 \times g$ for 5 min at room temperature. Take an aliquot for a cell count. Repeat this washing step to remove residual medium (including FBS proteins) and trypsin solution from the cell pellet.

3. Wash the cell pellets with sucrose buffer to reduce the salt content from the PBS wash solution in the final cell pellet.
4. Resuspend the cell pellet with 0.5 mL of lysis buffer such as 2D-DIGE lysis buffer. We have found that this 0.5 mL volume of lysis buffer is sufficient for optimal lysis of a 60–70% confluent T-75 cm^2 flask with a final cell count of approximately $4–5 \times 10^6$ cells per flask. However, this may require optimisation for individual cell lines.
5. Homogenise the lysate by gently passing it through a 21-gauge needle five times up and down using a 1 mL syringe. Avoid frothing the sample, as this can cause protein denaturation.
6. Incubate the lysate for 1 h at room temperature with gentle shaking on a rocking platform.
7. After incubation, centrifuge the lysates using a refrigerated microcentrifuge at $3,500 \times g$ for 15 min at 4°C. Transfer the middle layer of the supernatant containing extracted protein to a fresh chilled Eppendorf tube. For 2D-DIGE analysis, check the pH of the sample by spotting 3 μL onto a pH indicator paper to ensure that it lies between pH 8.0 and 9.0.
8. Divide the sample into smaller aliquots to reduce freeze–thaw steps and store at −80°C until use. Use a small aliquot to determine the protein concentration in the sample.
9. For protein quantification, we use Bradford assay that is compatible with urea and thiourea in the sample. Prepare dilutions of bovine serum albumin (BSA) stock for 0.125, 0.25, 0.5, and 1.0 mg/mL to generate a protein standard curve. Add 240 μL/well of Bradford protein assay reagent to the wells of a 96-well plate. Add 10 μL of protein standard dilution or sample (diluted 1:10) to the relevant wells of the 96-well plate. Assay all samples in triplicate. Incubate the samples for 5 min with gentle shaking on a rocking platform to ensure even mixing of sample with protein assay reagent. Read the absorbance of each standard and sample at 595 nm using a plate reader. The concentration of the protein samples is determined from the plot of the absorbance at 595 nm versus the concentration of the protein standard.

3.2. 2D Difference Gel Electrophoresis

3.2.1. Protein Labelling with Fluorescent Cy Dyes

1. To prepare the CyDye dyes for minimal DIGE protein labelling, thaw the 3 Cy dyes, Cy2, Cy3, and Cy5, from −20°C to room temperature for 5 min (see Note 5).
2. To each microfuge tube, add DMF to a concentration of 1 nmol/μL. Vortex each microfuge tube for 30 s to dissolve the dye. Briefly centrifuge the tubes for 30 s at $12,000 \times g$. Store the reconstituted dyes at −20°C for up to 3 months.

Table 1
Example of a simple 2D-DIGE experimental set up with four biological conditions (A–D) with four replicate samples per condition (1–4)

Gel no.	Cy2	Cy3	Cy5
1	Internal standard	A1	B1
2	Internal standard	A2	B2
3	Internal standard	B3	A3
4	Internal standard	B4	A4
5	Internal standard	D1	C1
6	Internal standard	D2	C2
7	Internal standard	C3	D3
8	Internal standard	C4	D4

3. Prepare a working dye solution of each Cy dye at a concentration of 400 pmol/μL by diluting the 1 nmol/mL stock with DMF. Vortex the dye solutions and briefly centrifuge at 12,000 × *g* in a microcentrifuge for 30 s to ensure all the dye solution is at the bottom of the tube. Store the working dye solution at −20°C for up to 1 week (see Note 6).
4. For protein sample labelling, add the relevant volume of sample equivalent to 50 μg of protein to labelled microfuge tubes. Prepare a Cy2 pool (internal standard) for each gel (50 μg) containing an equal concentration aliquot of each of the protein samples from all the samples used in the study. See Table 1 for an outline of a typical experimental set up.
5. Add 1 μL of each working dye solution (400 pmol/μL) to the 50 μg protein sample. Vortex each tube to ensure proper mixing followed by a brief centrifuge to ensure all contents are at the bottom of the tube. Leave samples to incubate on ice for 30 min in the dark by covering with tinfoil.
6. To stop the reaction, add 1 μL of 10 mM lysine to each tube. Vortex the samples to mix, followed by a brief centrifugation step, as before, and leave on ice for 10 min in the dark.
7. Mix the relevant Cy3- and Cy5-labelled samples that are to be run on the same gel, and add the relevant amount of Cy2 internal standard to each tube. Add an equal volume of 2× sample buffer to each tube and leave this mixture on ice for at least 10 min before loading onto rehydrated IPG strips for IEF.

3.2.2. First Dimension IEF

The IEF method described here is based on the IPGphor 3 IEF system (GE Healthcare) that we routinely use in our laboratory, and it is recommended to refer to the manufacturer's instruction manual for full details.

1. IPG strips contain dried acrylamide, and are stored long term until expiration date at –20°C. As a result, they need to be rehydrated before use. This is carried out in an Immobiline Dry Strip Reswelling tray. Slowly pipette 340 μL (the recommended volume for an 18 cm IPG strip, use other volumes for other length strips) of rehydration buffer into the centre of each slot. Remove any bubbles generated during this step. Remove the protective film (i.e. plastic backing) from the IPG strip, and position the strip gel side down and lower gently into the rehydration buffer. To ensure the entire strip is evenly coated with rehydration buffer, gently lift up and lower the strip using a forceps onto the entire surface of the solution avoiding trapping any bubbles. Overlay each strip with about 3 mL of IPG Cover Fluid starting on both ends of the strip and moving to the centre. Replace the protective lid onto the reswelling tray, and leave the strips to rehydrate overnight at room temperature for a minimum of 12 h.
2. For IEF using the IPGphor unit, the cup loading manifold is prepared for use. Place the manifold onto the IPGphor unit by inserting the "T" shape into the hollow provided. Using a forceps place the rehydrated strips gel side up in the correct orientation (+ to anode) and aligned just below the indented mark, to allow for the wicks to overlap the strip. Position the sample cups approximately 1 cm from the cathodic end of the strip and use the plastic insertion tool to secure the cups into place. Pipette approximately 25 μL of cover fluid into each sample cup in order to check for leaks. If no leaks are found, pipette 9 mL of cover fluid into each of the 12 lanes in the tray in order to completely cover the surface of the strips to prevent dehydration during IEF.
3. Place two paper wicks per strip on some tinfoil and add 150 μL of distilled water onto each wick to rehydrate them. Place the rehydrated wicks over both the cathodic and anodic ends of each of the strips, ensuring that approximately one third of the wick is positioned over the gel portion of the strip so as to guarantee a good contact with the electrodes. Position the electrodes so that they are in direct contact with where the wicks and the acrylamide on the strips overlap.
4. Just prior to loading the samples into the sample cups, briefly centrifuge the samples to remove any insoluble material. Load the samples into the sample cups with a pipette tip placed just beneath the surface of the cover fluid. The amount of protein

loaded per strip is 150 μg for 2D-DIGE, and approximately 400 μg to 1 mg of protein for preparative gels that are used for spot picking for protein identification.

5. Close the cover of the IPGphor unit and select the desired programme. Set the temperature at 20°C with a current of 50 μA per strip. We use the following IEF parameters for a typical 18 cm broad range strip. Step 1, 300 V for 3 h (step-and-hold); step 2, 600 V for 3 h (gradient); step 3, 1,000 V for 3 h (gradient); step 4, 8,000 V for 3 h (gradient); and step 5, 8,000 V for 4 h (step-and-hold) (see Note 7).
6. On completion of the IEF run, drain the strips of excess cover fluid and store in pyrex glass tubes at –80°C for up to 3 months or use directly in the second dimension.

3.2.3. Second Dimension SDS–PAGE

In our laboratory, we use the Ettan Dalt 12 gel for 2D-DIGE experiments; therefore, the method described here is based on this system, and it is again recommended to refer to the manufacturer's instruction manual for full details.

1. Use low fluorescent glass cassettes for DIGE experiments. Prior to casting gels, inspect the casting equipment and the glass cassettes to ensure they are clean. Assemble the cassettes in the gel caster so that the front and back plates are evenly aligned. Place a thin separator sheet in the gel caster unit followed by a glass cassette, and then alternate a thin separator sheet and a cassette until all 14 cassettes (including blank cassettes if needed) are in place. When the desired amount of cassettes is added, use the thicker separator sheets to bring the level so that it is even with the edge of the caster unit. Place the backing plate onto the caster frame and tighten the seal with the six screws provided.
2. Prepare a 12.5% acrylamide gel solution in a glass beaker. Immediately, prior to casting add the APS and TEMED. Pour the gel solution into the caster using a funnel taking care not to introduce any air bubbles into the system. Allow the gel solution to fill up to about 4 cm from the top of the smaller glass plate in the cassette. Add the displacement solution (coloured with bromophenol blue for easy visualisation) to the balance chamber until the bottom V-well of the gel caster is filled with this solution forcing the remaining gel solution into the gel caster. Immediately, overlay each of the gels with 1 mL of saturated butanol. Allow the gels to polymerise for at least 3 h (preferably overnight) at room temperature. Following this, gently unlock the caster, remove the gel cassettes, and rinse with distilled water. If the gels are not used immediately, they can be stored for up to 4 days in 375 mM Tris–Cl solution at 4°C.

3. Prepare the Ettan DALT-12 electrophoresis unit for SDS–PAGE by adding 6.48 L of distilled water and 720 mL of 10× SDS running buffer. Switch the pump on to cool the 1× SDS running buffer to 10°C.
4. The IPG strips require two equilibration steps prior to SDS–PAGE to saturate the proteins with SDS. Prepare SDS equilibration buffers A and B solutions prior to use. Add 10 mL of equilibration buffer A (with DTT) to each tube and incubate for 15 min with gentle agitation using an orbital shaker. DTT reduces the disulphide bonds of the proteins.
5. Remove equilibration buffer A and add 10 mL of equilibration solution B (with iodoacetamide). Incubate the strips for 15 min with gentle agitation. Iodoacetamide alkylates free thiol groups of the reduced –SH groups. During this equilibration step, prepare the 0.5% agarose overlay solution.
6. Using a forceps, rinse the IPG strips in 2× SDS electrophoresis running buffer and position between the two glass plates of the gel, with the plastic backing of the strip positioned against the larger glass plate. Push the strip down gently using a thin plastic spacer until it comes into contact with the surface of the SDS–PAGE gel. Ensure that the gel surface of the IPG strip does not touch the surface of the smaller glass plate. Pipette approximately 1 mL of 0.5% agarose solution over the IPG strips to seal in place.
7. Load the gel cassettes into the separation tank. Wet the outside of the glass cassettes with some distilled water, as this allows easier insertion into the slots in the tank. When all 12 slots in the upper chamber are filled, add approximately 2.5 L of 2× running buffer to the upper chamber. Place the cover on the unit and select the required running conditions. We usually use a run time of 18–24 h at 1.5 W per gel at 10°C, or until the bromophenol blue dye front reaches approximately 1 cm from the bottom of the gel.
8. When the run is completed, remove the gel cassettes from the tank using the DALT cassette removal tool and rinse with distilled water to remove excess running buffer. For DIGE gels, wrap the gel cassettes in tinfoil and leave at 4°C until ready for scanning.
9. For scanning DIGE-labelled samples, we use the Typhoon 9400 Variable Mode Imager. Switch the instrument on for at least 30 min prior to scanning. Select the appropriate emission filters and lasers for the individual Cy dyes [Cy2 520 BP40 Blue (488), Cy3 580 BP30 Green (532), and Cy5 670 BP30 Red (633)]. For each gel, it is important to do a pre-scan at a 1,000 μm resolution in order to obtain the optimal photo multiplier tube (PMT) voltage for sensitivity, and also to prevent

saturation of the signal from high abundant spots. A target maximum pixel value of 50,000–80,000 is usually suitable, and to achieve these values the PMT voltage may need to be adjusted (up or down) in small increments. Once the correct PMT voltage is found (PMT voltage ~ 500–600), scan the gel at 100 μm resolution, which results in the generation of three images, one each for Cy2, Cy3, and Cy5. The maximum pixel value should be similar for all gels in the experiment to allow for accurate quantitation of spot volumes.

10. Once the scanning is complete, import the gel images into analysis software such as Image Quant (GE Healthcare) in order to crop gels to facilitate spot matching in 2D gel differential analysis software.
11. Analyse the gels using 2D differential analysis software. There are a number of 2D differential analysis software packages available from different vendors that are suitable for the analysis of 2D-DIGE experiments [including Progenesis SameSpots (NonLinear Dynamics), Decyder (GE Healthcare)].
12. The 2D software calculates the consistency of the differences between samples across all gels and applies statistics to associate a level of confidence for each of the differences (e.g. student's *t*-test, ANOVA, etc.). An example of the output from such analysis is shown in Fig. 1. Designate differentially expressed proteins observed as "proteins of interest" and pick these protein spots from preparative 2D-PAGE gels for protein identification (see Note 8).

3.3. Sample Preparation for MS Analysis

1. Proteins separated by 2D-PAGE preparative gels for protein identification are normally visualised by staining with dyes such as CBB R250 and silver stain (gluteraldehyde-free), or fluorescence-based methods such as Sypro Ruby, Deep Purple, etc. Excise the protein(s) of interest from the gel (see Note 9), and transfer the gel spots to a 96-well plate or a PCR tube (see Note 10).
2. De-stain the gel spots using of a number of washes in a solution of 50% methanol in 40 mM ammonium bicarbonate for CBB-stained gels or a 1:1 mixture of 30 mM potassium ferricyanide:100 mM sodium thiosulphate for silver-stained gels. Destaining is not necessary for fluorescent-stained gel pieces.
3. Add neat acetonitrile to the destained gel spot to shrink and remove the liquid. Add enough trypsin solution to saturate the gel spot (see Note 11). Incubate at 4°C for 1 h. Add 20 μL of 20 mM ammonium bicarbonate solution to cover the gel spots and keep them wet during the overnight digestion. Incubate the gel spots overnight at 37°C (see Note 12).

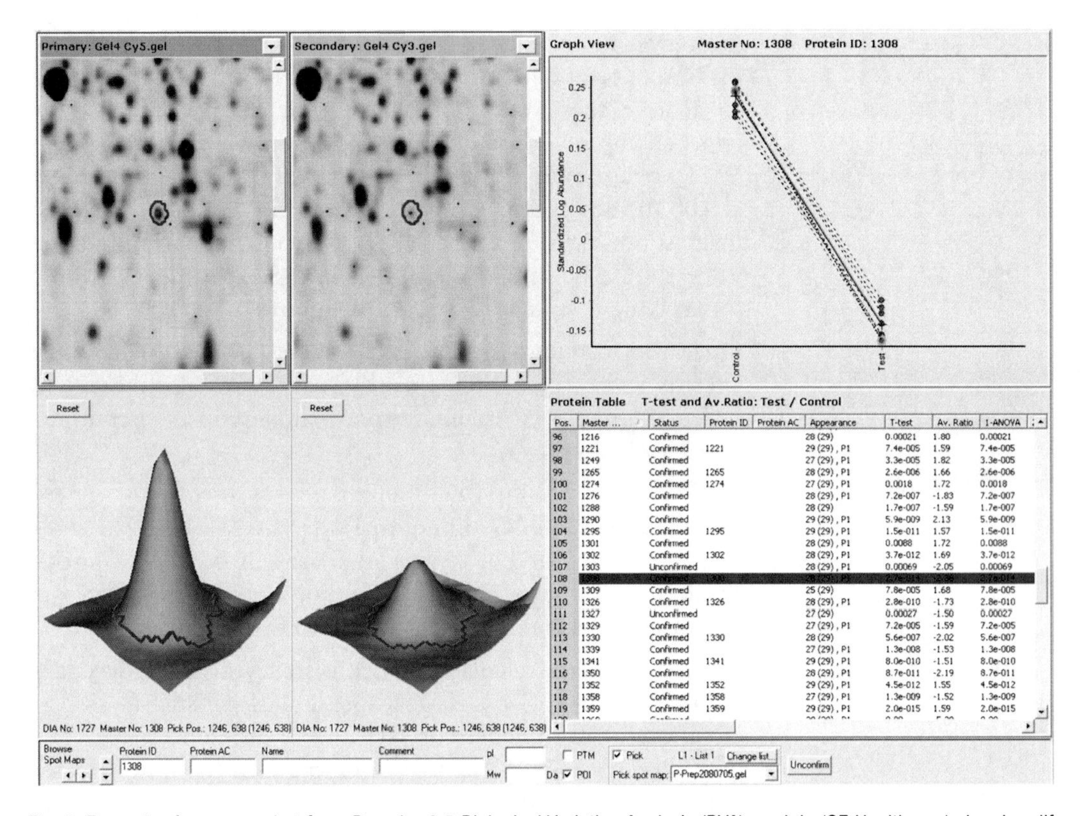

Fig. 1. Example of a screen shot from Decyder 6.5 Biological Variation Analysis (BVA) module (GE Healthcare) showing differential protein expression following 2D-DIGE of two recombinant Chinese hamster ovary cell lines, one a low producer of recombinant protein and the second a high producer of recombinant protein. A differentially regulated protein spot (no. 1308) that is 2.38-fold down-regulated ($p = 2 \times 10^{-14}$) is *highlighted*.

4. To extract the digested peptides from the gel pieces, add 20 μL of 50% acetonitrile/0.1% TFA solution to each piece and incubate for 15 min at room temperature on a shaker. Sonicate the gel plugs for 1 min in a sonicating water bath, and then transfer the extracted peptides to a clean tube. Repeat this step twice. Concentrate the extracted peptides using a speed vacuum centrifuge.
5. Add 20 μL of 0.1% TFA to the concentrated peptides. Samples can be analysed straight away or frozen at −20°C for future analysis.

3.4. LC–MS/MS for Protein Identification

1. Prior to analysis, equilibrate the columns in Solvent A for 10 min and set the column temperatures to 25°C using a column oven.
2. Inject 5 μL of digested protein samples using the injection pickup of the LC system onto a PepMap C18 trap cartridge (300 μm × 5 mm) at a flow rate of 25 μL/min for 5 min to desalt and concentrate the sample.

3. Elute the peptides from the trap column at a flow rate of 350 nL/min acetonitrile/water gradient (2–50% Solvent B in 30 min) onto a PepMap C18 capillary column (300 μm × 15 cm, 3-μm particles) directly into the electrospray tip. Peptides are eluted directly off the column into the LTQ Orbitrap XL mass spectrometer, which is the MS we use in our laboratory.
4. Re-equilibrate the columns in Solvent A for 10 min prior to analysis of the next sample.
5. The scan sequence of the MS is based on a data-dependent method. Acquire full scan in the Orbitrap at a resolution of 60,000 and then acquire subsequent MS/MS scans of the five most abundant peaks in the spectrum in the linear ion trap. Use dynamic exclusion to exclude multiple MS/MS of the same peptide. Set dynamic exclusion to a repeat count of 1, a repeat duration of 30 s and an exclusion list of 500. The general MS conditions we use are: electrospray voltage of 1.6 kV, ion transfer tube temperature of 200°C, collision gas pressure of 1.3 mTorr, normalised collision energy of 35% and an ion selection threshold of 500 counts for MS2. An activation *q*-value of 0.25 and an activation time of 30 ms for MS2 acquisitions are also used.
6. Identify the peptides using the tandem mass spectrum (MS/MS) data generated by mass spectrometry. Search each tandem mass spectrum against a database using database-searching software packages (see Note 13). Peptide identifications are reported in terms or XCorrelation scores and probability scores, as in the case for the SEQUEST algorithm (13) that we use in our analysis (Thermo Fisher Scientific). If several statistically significant peptides are identified from the same protein, then generally this protein identification is accepted.

4. Notes

1. All chemicals must be of the highest purity. All solvents must be of the highest purity and LC–MS grade. Water for LC–MS analysis must also be LC–MS grade.
2. When thawing an aliquot of 2D lysis buffer, vortex briefly to ensure that all the urea and thiourea goes back into solution. It is also important to avoid heating the lysis buffer above room temperature as this can result in the degradation of urea to isocyanate leading to the carbamylation of proteins.
3. A new bottle of DMF should be opened every 3 months, as once the bottle is opened it will start to degrade generating amine compounds, which can affect the efficiency of the labelling reaction.

4. There are many commercially available stains for visualising proteins on 2D gels. For mass spectrometry analysis, the silver stain method should not use glutaraldehyde.
5. 2D-DIGE is a relatively expensive technique for the proteomic analysis of biological samples. Therefore, it is important to ensure that the experiment is well designed with enough biological replicate samples to ensure that the experiment is a success. In our laboratory, we use the Cy2 internal standard, and also differentially label one half of the biological replicate samples with Cy3 and the other half with Cy5 (irrespective of experimental condition such as "Normal" or "Diseased") to rule out dye bias.
6. It is possible to use lower amounts of Cy dye per 50 μg of protein. We routinely use a 200 pmol/50 μg of protein without loss in sensitivity of protein detection, rather than the 400 pmol per sample as recommended by the manufacturer (GE Healthcare). This significantly reduces the cost of the experiment.
7. The total volthours for an IEF experiment using broad range strips should be in the range of 40–50 kVh. The number of volthours may need to be increased when using narrower range strips or when loading higher amounts of proteins such as used in preparative gels.
8. If gels are to be used for "spot picking" the plates need to be treated with bind-silane solution to stick the acrylamide gel to the plates to allow for easier spot picking, especially if using a spot picker robot.
9. When working with SDS–PAGE gels, gloves must be worn at all times to minimise keratin and dust contamination which can affect your LC–MS results. When excising protein from gels use sterile scalpel blades, and this work should be done in a laminar flow cabinet to minimise the possibility of any hair, dust or skin flakes contaminating the sample with unwanted keratins.
10. If you are using 96-well plates to carry out protein digests prior to LC ensure that you use polypropanol-grade plastic, and not regular polystyrene plastic plates. The use of polypropylene tubes and tips helps to minimise protein loss by adsorption.
11. The most common enzyme used for peptide generation for LC–MS analysis is trypsin, which has been modified by reductive alkylation to suppress trypsin autolysis. We use modified porcine trypsin (Promega) in our laboratory.
12. There are enzymes other than trypsin that can be used for protein digestion, which have different cleavage specificities, such as lysyl endopeptidase (Lys-C) and chymotrypsin.
13. We use the SEQUEST (Thermo Fisher Scientific) search engine for protein identification. There are other similar search engines including MASCOT (Matrix Science) and X!Tandem.

References

1. Hanash, S. (2003) Disease proteomics. *Nature* **422**, 226–232.
2. Görg, A., Weiss, W., and Dunn, W.J. (2004) Current two-dimensional electrophoresis technology for proteomics. *Proteomics* **4**, 3665–3685.
3. Görg, A., Drews, O., Lück, C., Weiland, F., and Weiss, W. (2009) 2-DE with IPGs. *Electrophoresis.* **30**, Suppl 1:S122–32.
4. Smith, R. (2009) Two-Dimensional Electrophoresis: An Overview. *Methods Mol. Biol.* **519**, 1–16.
5. Unlü, M., Morgan, M.E., and Minden, J.S. (1997) Difference gel electrophoresis: a single gel method for detecting changes in protein extracts. *Electrophoresis* **18**, 2071–2077.
6. Alban, A., David, S.O., Bjorkesten, L., Andersson, C., Sloge, E., Lewis, S., and Currie, I. (2003) A novel experimental design for comparative two-dimensional analysis: two-dimensional difference gel electrophoresis incorporating a pooled internal standard. *Proteomics* **3**, 36–44.
7. Mann, M., Hendrickson, R.C., and Pandey, A. (2001) Analysis of proteins and proteomes by mass spectrometry. *Annu. Rev. Biochem.* **70**, 437–473.
8. Corona, G., De Lorenzo, E., Elia, C., Simula, M.P., Avellini, C., Baccarani, U., Lupo, F., Tiribelli, C., Colombatti, A., and Toffoli, G. (2010) Differential proteomic analysis of hepatocellular carcinoma. *Int. J. Oncol.* **36**, 93–9.
9. Dowling, P., O' Driscoll, L., Meleady, P., Henry, M., Roy, S., Ballot, J., Moriarty, M., Crown, J., and Clynes, M. (2007) Two-dimensional difference gel electrophoresis of the lung squamous cell carcinoma versus normal sera demonstrates consistent alterations in the levels of 10 specific proteins. *Electrophoresis* **28**, 4302–10.
10. Dowling, P., Wormald, R., Meleady, P., Henry, M., Curran, A., and Clynes, M. (2008) Analysis of the saliva proteome from patients with head and neck squamous cell carcinoma reveals differences in abundance levels of proteins associated with tumour progression and metastasis. *J. Proteomics* **71**, 168–75.
11. Keenan, J., Murphy, L., Henry, M., Meleady, P., and Clynes, M (2009) Proteomic analysis of multidrug-resistance mechanisms in adriamycin-resistant variants of DLKP, a squamous lung cancer cell line. *Proteomics* **9**, 1556–66.
12. Lau, T.Y., Power, K.A., Dijon, S., de Gardelle, I., McDonnell, S., Duffy, M., Pennington, S., and Gallagher, W.M. (2010) Prioritization of candidate protein biomarkers from an in vitro model system of breast tumor progression towards clinical verification. *J. Proteome Res.* **9**, 1450–9.
13. MacCoss, M.J., Wu, C.C., and Yates, J.R. 3rd. (2002) Probability-based validation of protein identifications using a modified SEQUEST algorithm. *Anal. Chem.* **74**, 5593–5599.

Chapter 10

Design, Construction, and Analysis of Cell Line Arrays and Tissue Microarrays for Gene Expression Analysis

Kathy Gately, Keith Kerr, and Ken O'Byrne

Abstract

Cell line array (CMA) and tissue microarray (TMA) technologies are high-throughput methods for analysing both the abundance and distribution of gene expression in a panel of cell lines or multiple tissue specimens in an efficient and cost-effective manner. The process is based on Kononen's method of extracting a cylindrical core of paraffin-embedded donor tissue and inserting it into a recipient paraffin block. Donor tissue from surgically resected paraffin-embedded tissue blocks, frozen needle biopsies or cell line pellets can all be arrayed in the recipient block. The representative area of interest is identified and circled on a haematoxylin and eosin (H&E)-stained section of the donor block. Using a predesigned map showing a precise spacing pattern, a high density array of up to 1,000 cores of cell pellets and/or donor tissue can be embedded into the recipient block using a tissue arrayer from Beecher Instruments. Depending on the depth of the cell line/tissue removed from the donor block 100–300 consecutive sections can be cut from each CMA/TMA block. Sections can be stained for in situ detection of protein, DNA or RNA targets using immunohistochemistry (IHC), fluorescent in situ hybridisation (FISH) or mRNA in situ hybridisation (RNA-ISH), respectively. This chapter provides detailed methods for CMA/TMA design, construction and analysis with in-depth notes on all technical aspects including tips to deal with common pitfalls the user may encounter.

Key words: Cell line array, Tissue microarray, Immunohistochemistry, Tissue arrayer, Image acquisition system

1. Introduction

The concept of multitissue blocks was first introduced in 1986 when Battifora described the "sausage" or multitumour tissue block (1). Kononen et al. (2) developed this idea further with the introduction of the high precision punching instrument, enabling the precise relocation of tissue cylinders from a formalin-fixed paraffin-embedded "donor" block into a "recipient" paraffin block.

Lorraine O'Driscoll (ed.), *Gene Expression Profiling: Methods and Protocols*, Methods in Molecular Biology, vol. 784, DOI 10.1007/978-1-61779-289-2_10, © Springer Science+Business Media, LLC 2011

Beecher Instruments pioneered the arrayer technology, but several companies now supply both manual and automated arrayers. Unitma (Seoul, Korea) have recently introduced their Quick-Ray™ system, whilst Veridiam (Oceanside, Calif) have developed a semi-automated system with an integrated microscope station. Recent advances in digital pathology allow TMAs to be digitally scanned and automatically analysed, eliminating the tedious process of manual evaluation. Complex image analysis algorithms assist pathologists with the interpretation of IHC staining and subcellular localisation of protein markers. TMA technology is extremely powerful, providing the researcher with the potential to derive extensive gene expression profiles that are invaluable, particularly in the areas of tumour biology, clinical oncology and the development of diagnostic tests (3–5). TMA technology also plays an important role in the education of pathologists, quality control in immunohistochemistry (IHC), and tissue banking. In recent years, the number of studies, including data generated by TMA experiments, has increased logarithmically. Several reviews (6–10) and book chapters (11, 12) provide detailed descriptions of different aspects of tissue microarray technology.

The generation of Cell line arrays (CMAs) and tissue microarrays (TMAs) is a lengthy process that involves a series of steps including (1) design, (2) construction, and (3) analysis. Each of these steps can be further divided into subheadings, each of which will be examined in detail in Subheading 3.

2. Materials

1. Phosphate-buffered saline.
2. Trypsin/EDTA.
3. 10% Neutral-buffered formalin (10% NBF).
4. UltraPure™ Low Melting Point (LMP) Agarose.
5. Haematoxylin.
6. Eosin.
7. Paraplast X-Tra paraffin.
8. Stainless steel deep moulds.
9. Tissue-embedding cassettes.
10. SuperFrost® Plus slides.
11. Coverslips.
12. Manual tissue arrayer (Beecher Instruments).
13. A pair of punches with stylets; 0.6, 1.0, 1.5 or 2.0 mm.

14. Laboratory oven.
15. Feather S35 microtome blades.
16. Semi-automated rotary microtome.
17. Floatation Water Bath.
18. Slide Drying Hotplate.
19. Histology pen.

3. Methods

The methods described below are based on the manual tissue arrayer (Beecher Instruments), but will also provide the users of automated tissue arrayers with a good insight into the technology. Detailed methods of individual automated tissue arrayers can be found in their instruction manuals.

3.1. CMA/TMA Design

3.1.1. Type of Array

Careful consideration must be given to the proposed use of the array before the design process begins. CMAs are generated to use in the optimisation of antibodies for IHC analysis, or fluorescent probes for fluorescent in situ hybridisation (FISH). A panel of cell lines are chosen that either (1) do not express or (2) over-express the target of interest, automatically providing an internal negative and positive control (see Note 1). TMAs with a small number of different tissue types obtained from discarded surgical specimens, e.g. appendix, thyroid, and liver can also be generated for use in the working up of new antibodies or probes and for the identification of an appropriate tissue to use as a positive/negative control in larger TMAs.

Predictive arrays can be used to optimise a new antibody to determine whether it can be used as a suitable marker of disease, often malignancy. A TMA with matched normal and tumour tissue from a small cohort of individuals (20–30 patients) with a specific disease, e.g. non-small cell lung cancer (NSCLC) could be generated to use in this way. These arrays may also be used to test for variations between batches of antibodies (see Notes 2 and 3) prior to testing on larger TMAs.

Prognostic studies require a large number of cases that have long-term survival data in order to have the appropriate statistical power. The generation of a TMA set that represents a large cohort of individuals (several hundred) within several blocks is a very valuable, but limited resource for such studies. These prognostic TMAs can be used in multiple studies to answer several different scientific questions, and allow the researcher to build up an extensive profile of a disease. The steps outlined below are extremely important to follow when generating large array sets.

3.1.2. Preparation of Cell Line Donor Blocks for CMAs

1. Grow suspension/adherent cell lines in a T75 flask until they reach 60–80% confluence.
2. Spin down suspension cells and wash pellet twice with 1× PBS. Remove media from adherent cells, detach with Trypsin/EDTA (5 min, 37°C), and wash pellet twice with 1×PBS.
3. Resuspend the cell pellet in 10% (v/v) NBF in a 1.5-ml tube and incubate at 4°C for 12–24 h or until further processing.
4. Prepare 10 ml of 1% molten LMP agarose (in 1× PBS) in a microwave. Incubate the agarose in an oven at 55°C, while carrying out subsequent steps.
5. Using scissors or a razor blade, cut the tapered end off a 1 ml syringe.
6. Centrifuge cells at 150×*g* for 5 min, gently remove supernatant, wash once in 1× PBS, and reform the pellet.
7. Add ~0.5 ml of the molten agarose to the cell pellet and resuspend cells *(*see Note 4*)*.
8. Draw the cell/agarose suspension into the barrel of the cut 1-ml syringe, using the plunger.
9. Allow the cell/agarose suspension to solidify in the barrel of the syringe for at least 30 min at 4°C.
10. Eject the cell plug by pushing down slowly on the plunger. The plug can be cut in half to generate two cell blocks.
11. The cell/agarose plugs can then be embedded in paraffin blocks using standard processing methods.

3.1.3. Donor Tissue Selection

The success of a TMA depends on the careful selection of the donor tissue. Factors we consider when constructing our NSCLC TMAs are the following: (1) inclusion of matched normal/tumour tissue for each donor; (2) what areas of tumour are of interest, e.g. leading edge, vessels, etc.; (3) are the areas of stroma around the tumour also of interest; and (4) separate different histologies (adenocarcinomas vs. squamous cell carcinomas) into different TMA blocks (see Note 5). Once these factors have been decided, the appropriate tissue blocks are identified and acquired from the pathology archives. This step is the most time-consuming part of the entire TMA process.

Preparation of Tissue Donor Blocks for TMAs

1. Initially, the histology database is accessed and details regarding each block are determined. These include whether the block contains tumour, normal tissue, etc.
2. Donor blocks are retrieved from the pathology archives and a fresh 5-μm section is cut from each block and stained with haematoxylin and eosin (H&E) *(*see Notes 6 and 7*)*.
3. The H&E-stained slide is reviewed by a pathologist who identifies the area of interest (tumour, leading edge of tumour,

normal, etc.) from which the core can be extracted. The representative regions are circled with a histology pen.

4. Each region can be assigned its own unique identifier (number and letter), such that it can be easily related to a specific location within a specific block.
5. Overlay the circled H&E glass slide over the corresponding area of interest on the donor block.
6. Arrange all block and slide pairs in the same order that they appear on the array map (see Subheading 3.1.5).

3.1.4. Core Size and Number

The number of cell line pellets to be embedded in the CMA, the availability and different types of donor tissue to be included in the TMA all determine the size and quantity of cores to be used. The range of punch sizes available are 0.6, 1, 1.5, and 2 mm. Initially, there was concern whether a 0.6-mm core is sufficient to accurately represent the heterogeneous morphology of tumour tissue. Between 1998 and 2003, several groups carried out studies to validate the use of TMAs in cancer research by comparing the results of TMA analysis with large tissue sections in different cancer types (13–16). These studies have shown that two to four 0.6-mm cores yield the same information as the larger cores. Although there is no standardised sampling method, taking multiple cores from several different areas of a tumour block is the best way to achieve a high degree of concordance between results for full sections and TMA cores. Multiple cores provide a broader representation of the tumour, particularly where there is a high degree of heterogeneity. This is of particular importance when examining protein markers that are expressed focally as opposed to homogeneously, e.g. CAIX (a marker of hypoxia). When using larger cores the 1- or 1.5-mm punches are recommended, as it has been shown that the 2-mm punch can cause damage to both donor and recipient blocks (see Note 8). Multiple cores also allow for any loss or damage of individual cores that can occur during the processing. The total number of cores to be incorporated into a single TMA block also needs consideration, and this will depend on both the size of the cores and the size of the recipient block. Up to 1,000 cores of 0.6-mm diameter can be placed in a TMA block, however 600 cores is usually the limit (see Note 9). The spacing between two adjacent cores on an array will vary according to punch size. Spacing of 0.2 mm between individual 0.6-mm diameter cores is standard, however a slightly larger space can also be used. Table 1 shows the standard spacing between different size cores in a recipient block without having to remove the recipient block from the holder and rotate it during the construction of the array.

3.1.5. CMA/TMA Map

Using a Microsoft Excel database or dedicated TMA software, a grid map is designed specifying the exact location of each core on

Table 1
Core spacing and number of cores using different punch sizes

0.6 mm	0.2 mm	20 × 20 cores	400
1.0 mm	0.3 mm	16 × 13 cores	208
1.5 mm	0.4 mm	11 × 9 cores	99
2.0 mm	0.5 mm	9 × 6 cores	54

	A	B	C	D	E	F	G	H	I	J	K	L	M
1	Lung Array 1		Sector A										
2				1	2	3	4	5	6	7	8	9	
3	Liver	A1-A3	A	Liver	Liver	Liver	LN-02.001	LN-02.001	LN-02.001	LT-02.001	LT-02.001	LT-02.001	
4	LN-02.001	A4-A6	B	LN-02.002	LN-02.002	LN-02.002	LT-02.002	LT-02.002	LT-02.002	LN-02.003	LN-02.003	LN-02.003	
5	LT-02.001	A7-A9	C	LT-02.003	LT-02.003	LT-02.003	LN-02.004	LN-02.004	LN-02.004	LT-02.004	LT-02.004	LT-02.004	
6	LN-02.002	B1-B3	D	LN-02.005	LN-02.005	LN-02.005	LT-02.005	LT-02.005	LT-02.005	LN-02.006	LN-02.006	LN-02.006	
7	LT-02.002	B4-B6	E	LT-02.006	LT-02.006	LT-02.006	LN-02.007	LN-02.007	LN-02.007	LT-02.007	LT-02.007	LT-02.007	
8	LN-02.003	B7-B9	F	LN-02.008	LN-02.008	LN-02.008	LT-02.008	LT-02.008	LT-02.008				
9	LT-02.003	C1-C3											
10	LN-02.004	C4-C6											
11	LT-02.004	C7-C9											
12	LN-02.005	D1-D3		A1	A2	A3	A4	A5	A6	A7	A8	A9	
13	LT-02.005	D4-D6		B1	B2	B3	B4	B5	B6	B7	B8	B9	
14	LN-02.006	D7-D9		C1	C2	C3	C4	C5	C6	C7	C8	C9	
15	LT-02.006	E1-E3		D1	D2	D3	D4	D5	D6	D7	D8	D9	
16	LN-02.007	E4-E6		E1	E2	E3	E4	E5	E6	E7	E8	E9	
17	LT-02.007	E7-E9		F1	F2	F3	F4	F5	F6				
18	LN-02.008	F1-F3											
19	LT-02.008	F4-F6											

Fig. 1. Example of a TMA map used to guide the construction and the subsequent scoring of the array.

the CMA/TMA. The map is used to guide the array construction and subsequent scoring and can be linked to database files that contain IHC images, staining scores, and the clinicopathological data on each case. Figure 1 is an example of a map we use to design and construct our NSCLC arrays. There are nine cores along the *x*-axis and six down the *y*-axis. The first three cores on the top left corner of the array (A1–A3) are control cores. In this example we have used liver tissue, however any normal tissue and/or cell lines can be used (see Note 10). The next six cores are from the same patient (who has been given the unique identifier "001"). The first three cores (A4–A6) represent normal lung tissue and the next three cores (A7–A9) represent tumour tissue. Each set of triplicate cores is colour coded to allow an individual to be easily identified. This database can then be linked to files that contain staining data, etc. An example of how to create an interactive database for image viewing and data input can be found at: http://icg.cpmc.columbia.edu/cattoretti/Protocol/Immunohistochemistry/TissueArray.html.

It is a good idea not to design a completely symmetric array. The last row of this array is shorter than the other rows and determines the orientation of the array when mounting it onto a glass slide and during the scoring of the sections.

3.2. CMA/TMA Construction

The CMA/TMA is constructed using a manual or automated tissue arrayer. It is necessary for the beginner to familiarise themselves with both the equipment and the coring technique by preparing several practise arrays, before attempting to use precious biopsy/ resected tissue that is destined for the large TMA set (see Notes 11 and 12). When possible, construct multiple replicas of each TMA simultaneously (see Note 13). Detailed methods for array construction using the manual arrayer (Beecher) are outlined below.

3.2.1. Preparation of Recipient Block

1. Melt soft paraffin, such as Paraplast X-Tra, at 56–58°C, and then pour into a deep mould (5–10 mm). Place a standard tissue cassette on top of the melted paraffin wax (see Notes 14 and 15).
2. Allow the block to cool at room temperature. Once cool, separate the cassette and mould.
3. Ensure that the block surface is flat (see Note 16).

3.2.2. Tissue Arrayer Set up

1. A "test" paraffin block is inserted into the magnetic paraffin block holder on the base plate of the arrayer. The block is tightened into position with the aid of two small screws.
2. Instal a pair of punches into the arrayer (see Note 17). A right-handed operator will place the larger punch (donor) in the right punch holder allowing the donor block to be held in place with the left hand. The smaller punch (recipient) is placed in the left punch holder (a left-handed operator will use the opposite configuration).
3. Using the *X–Y* precision guides, the block is lined up so that the hollow needle of the recipient block is above the position of the first punch, usually at the top left corner of the block (see Note 18). The position of the punch over the block can be assessed by gently lowering the punch, making a superficial mark on the surface of the paraffin block.
4. Next, move the turret to switch the donor punch into position and again make a mark in the paraffin. Ensure both marks overlap precisely (see Note 19).
5. Remove the "test" paraffin block and insert the recipient para ffin block into the holder.
6. Zero the *X–Y* micrometres.

3.2.3. Construction of CMA/TMA Blocks

1. It may be necessary to adjust the depth stop by tightening or loosening the nut at the top left of the turret. This will determine

the depth of the core that will be removed from the recipient block by controlling the downward movement of the turret. The "Depth Stop Kit" by Beecher can also be used to control the length of the donor tissue core (see Note 20).

2. Ensuring the recipient punch is in place, gently press down the top of the turret to bring the needle downwards towards the recipient block and push it into the paraffin.
3. When the depth stop blocks the downward motion, rotate the handle of the punch to the left and then back to the right, while still pressing down on the turret top. This movement helps free the core from the recipient paraffin block (see Notes 21 and 22).
4. Remove the pressure on the turret, allowing the springs to pull the punch back up to its original position.
5. Eject the paraffin core from the punch by pressing down on the stylus and check the length of the core (see Note 23).
6. Move the turret to the right to allow the donor needle to be brought into the correct sampling position.
7. The donor block bridge is then placed in position above the recipient block. The donor block/slide pair is placed on top of the donor block bridge.
8. Gently guide the turret downwards towards the slide, and ensure it comes to rest in the centre of the circled region of interest.
9. Carefully remove the slide from the donor block ensuring the block remains in position.
10. The needle is then pushed into the region containing the cell pellet or region of interest on the donor block.
11. Rotate the handle of the punch to the left and then back to the right, while maintaining pressure on the turret top to ensure removal of the core.
12. Release the pressure on the turret, allowing the springs to raise it to its resting position.
13. Carefully remove the donor block and its bridge allowing the recipient block to be exposed. Slowly lower the needle containing the core towards the empty hole in the recipient block. When it is just above the hole, carefully push the core from the needle and deposit it into the hole (see Notes 24 and 25).
14. Move the recipient block to the next *X–Y* position using the *x*-precision guide. When using the 1.0-mm punch, spacing of 1.3 mm is used between core centres (see Table 1). Repeat steps 2–14 until the last core in the row is in place (see Note 26).
15. Using the *x*-axis precision guide go back to position zero and move the *y*-axis knob to 1.3 mm. Repeat steps 2–16 until the block is complete.

16. When all the cores are in place, remove the block from the base plate and place face down on a clean glass slide. Incubate the block/slide at 37°C for 15 min to ensure the cores adhere to the walls of the cored recipient block.
17. Remove the block/slide from the incubator and while it is still warm and elastic, carefully press downward until all the specimens are at an even level.
18. Let the TMA block cool on ice before removing the slide.
19. The TMA block is now ready for sectioning.

3.2.4. CMA/TMA Sectioning

Sectioning should be carried out by a highly experienced histotechnologist, as TMAs are often difficult to section, particularly if multiple tissue types are included in the array. Using a dedicated microtome for all steps of the construction process is recommended (see Note 16). A series of at least 20 sections should be cut from the TMAs at the same time in order to minimise tissue wastage. To improve the section quality the "Paraffin tape-transfer system" by Instrumedics Inc. can be used (see Note 27). If using conventional sectioning, briefly heat the block to 37°C for about 20–30 min and allow it to cool back to room temperature before sectioning. Ensure the blades are changed regularly. Stain the first section and every 50th section of the TMA block with H&E to monitor specimen morphology and representativity. To preserve antigenicity, the sections can be stored at –20°C or paraffin coated and stored in liquid nitrogen (17).

1. Set the waterbath to 37°C.
2. Gently face off the TMA block on a dedicated microtome.
3. Cut sections of 4–5 μm from the TMA block. Always place sections on glass slides in the same orientation.
4. Dry the slides at 50°C overnight in a vertical position.
5. Place the slides in a slide box, label, and store at –20°C or, for long-term storage, coat in paraffin and store in liquid nitrogen (see Note 28).

3.2.5. CMA/TMA Slide Staining

IHC can be performed on TMA slides with minimal changes to standard protocols. Harsher enzymatic digestions are needed for FISH, which may cause some tissue detachment. Multiple cores will partially compensate for this problem.

3.3. CMA/TMA Analysis

TMA slides can simply be viewed under a microscope, and the observations inputted into a spreadsheet by a pathologist or trained scientist. However, a series of TMA slides produces vast amounts of data including images and quantified biomarker staining results linked to patient clinicopathological characteristics. Therefore, it is extremely important that these data are managed effectively in order to both preserve them securely and also to ensure that they can be manipulated and evaluated by the researcher, allowing as much information as possible to be retrieved. To this end, TMA slides are usually digitally scanned using an automated imaging acquisition system that allows the researcher to either score the images in silico, or to have the system perform automated image analysis. Several imaging and visualisation systems are now available including the Scanscope system by Aperio Technologies (Vista, CA), TMAx by Beecher Instruments (Sun Prairie, WI), BLISS system by Bacus Laboratory (Lombard, IL), Ariol by Applied Imaging (San Jose, CA), AQUA software by HistoRx (New Haven, CT), and Medscan by Trestle (Irvine, CA). We use the Scanscope system, together with the TMALab Microarray Analysis Tool, which allows the scanning of images, image analysis, data management, and export of data and images to database or text files. These files can then be used for hierarchical clustering or other statistical analysis. Figure 2 is an example of how MatLab 7 (MathWorks, Apple Hill Drive, MA, USA) software was used with Scanscope images (Aperio CA, USA) to develop an automated image analysis algorithm to identify and quantify positive nuclear IHC staining in tumour cells. This novel approach provides a useful tool for the quantification of biomarkers on TMA sections, as well as for objective identification of appropriate cut-off thresholds for biomarker positivity (18). Open-source softwares for constructing the array, digital scanning, and image analysis are available online at John Hopkins University (http://tmaj.pathology.jhmi.edu), Stanford University (http://genome-www.stanford.edu/TMA/), and the University of Leeds (http://www.bioinformatics.leeds.ac.uk/tmadb).

4. Notes

1. Sometimes it may be necessary to activate a specific pathway in order to over-express a target of interest, e.g. EGF can be added to a cell line in culture to activate the EGFR pathway. An extra flask of cells can also be grown in order to generate cell lysates, which can be used to quality control antibodies by Western blot analysis.
2. When generating a test array of tumours, it may be advantageous, depending on the research to be undertaken, to use tumours reflecting the expected range of different histotypes (e.g. adenocarcinoma, squamous cell, etc.) and stages of the disease (e.g. stage I–IV). This may have a considerable impact on the results obtained.
3. We often find variability of staining between different lot numbers of the same antibody. Therefore, it may be necessary to re-optimise antibody concentrations. Having a test TMA (see above) is important, as it preserves the larger TMAs.

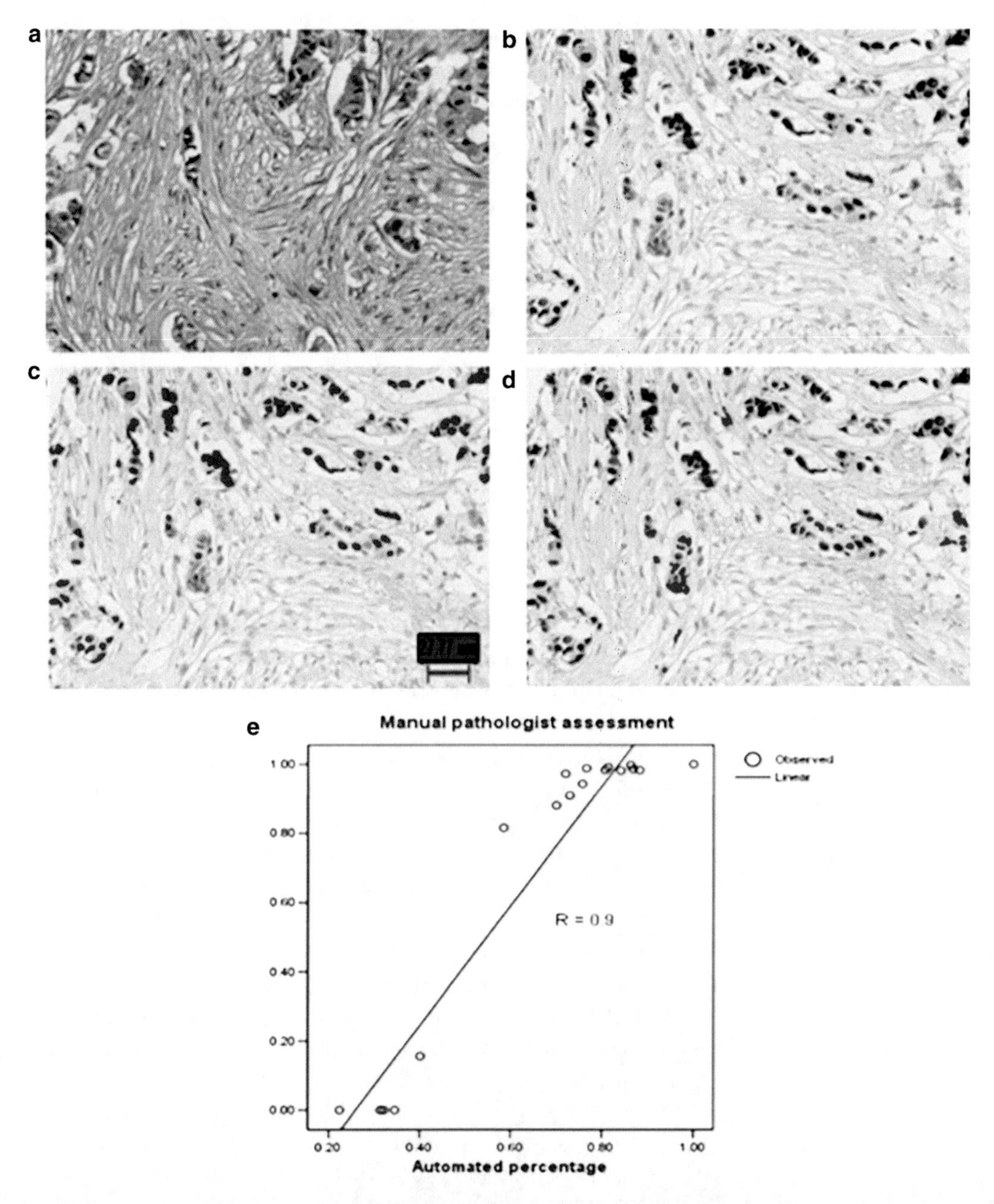

Fig. 2. Overview of the automated image analysis process: the stepwise image process underlying a nuclear algorithm is demonstrated using representative digital slides. (**a**) H&E-stained breast tumour section. (**b**) Original IHC section after the extraction of DAB-positive tumour nuclei. (**c**) Identification in *red* of DAB-positive tumour nuclei. (**d**) Identification in *blue* of DAB-negative tumour nuclei. (**e**) *Scatter plot* demonstrating strong correlation between automated scores and manual annotation of the same cores by a pathologist.

4. Steps 7 and 8 should be carried out quickly as the agarose will begin to solidify at room temperature.
5. When using TMAs to correlate staining with clinical outcome, it is important to try and use donors who have a minimum of 5-year survival data. It is also necessary to exclude tissue from

donors who died <30 days after surgery. Patients whose tumours were incompletely excised may not be suitable for inclusion; at least in the data set, if not in the TMA itself.

6. A H&E-stained slide from the pathology archive can also be used, provided it is representative of the block, i.e. not too many sections were cut from the block, since the H&E slide was prepared.
7. Archival blocks as old as 40 years have been successfully used in TMAs. Care should be taken to ensure that older archival tissues were fixed in a way which will not compromise antigenicity or nucleic acid integrity. The depth of the donor block will dictate how many consecutive sections can be cut from the TMA. A minimum of 1 mm is necessary, but ideally a depth of 3–4 mm is used. If the blocks do not contain enough tissue, one must then decide whether it would be more beneficial to cut full face sections and stain standard slides instead of preparing TMAs and risk wasting valuable tissue.
8. Allow donor blocks to warm up under a low wattage bulb so that they will become softer, easier to punch, and less likely to crack. When using larger cores ensure that enough space is left between the cores, as this will minimise the chances of the donor block buckling (see Table 1).
9. $1,000 \times 0.6$-mm cores can be arrayed in a (40×25 mm) recipient block, however constructing blocks with over 600 cores requires very high density and fidelity that may be difficult to maintain. It also requires rotating the recipient block 180° during the arraying process, as the *X*–*Y* precision guide rails are limited to moving ~25 mm in each dimension. It may be difficult to re-align the array after rotation.
10. It is important to integrate a few cores of control tissue at designated locations along the array. These controls can act as internal positive/negative controls for immunohistochemical analysis, as a guide to orientate the array, and also as a quality control check within the array. When scoring very large arrays, it can be reassuring to find control tissues at the correct (*X*, *Y*) position as it is an indication that the array has been accurately constructed.
11. For the inexperienced operator, it is highly recommended that a few practise blocks are made, using the 1.5- or 2-mm punch and utilising some excess tissue or cell line donor blocks, before actual TMA construction is undertaken. Initially, the larger punches are generally easier to work with.
12. It may be easier to construct the array using a magnifier with an attached light, particularly if the tissue arrayer is located in an area with low ambient lighting.

13. The design step of an arraying project is time consuming, therefore it is a good idea to construct multiple replicas of the array blocks simultaneously. The generation of replicate TMAs is faster and more convenient with the automated arrayer.
14. Recipient blocks are prepared using Paraplast X-tra, as it is softer than standard paraffin and easier to core. Use deep moulds and avoid trapping air bubbles under the cassette as they can result in holes in the paraffin.
15. Although conventional histological cassettes are routinely used to make donor paraffin blocks, these are most suitable for use with 0.6-mm cores. An individual formed paraffin wax block is superior for 2-mm cores, as the tension within the paraffin wax can result in the array breaking if a conventional cassette is used.
16. Use a dedicated rotary microtome to face off the recipient block. This will ensure that the block face is smooth and minimises the amount of block realignment that will be needed during sectioning of the TMA. Ensure the back of the block is also level.
17. Care is needed when tightening screws on the tissue arrayer as overtightening can damage the components and misalign the punches.
18. To prevent the recipient block from cracking, it is important to allow a margin of 2.5 mm of paraffin around the entire array.
19. If the marks do not coincide, then adjustment is needed. Punch holes appearing one in front of the other need front-to-back adjustment. Holes that appear side by side require left-to-right adjustment. To adjust the position of the punches in the holder, loosen the four screws holding the v-block and use the smallest hex key to advance the setscrew by turning it clockwise or anticlockwise to move it backward. Left-to-right adjustment is achieved by using the setscrews on the left and right sides of the turret.
20. The Depth Stop Kit (Beecher) includes preset stops in 1-mm increments (from 2 to 6 mm), enabling control of the length of the tissue core and a companion set of depth stops for holding the donor punch just above the recipient block when depositing tissue. The kit allows more consistent results and less waste.
21. Rotating the stylus back and forth twice will usually free the core from the block.
22. The punches can be sharpened if they become dull (only after thousands of punches).
23. Remove any excess paraffin that is retained on the stylus/ punch complex by pressing down on the stylus and wiping it

with a Kimwipe or similar to remove any excess. The stylus can also be removed and wiped clean.

24. Do not push the donor punch all the way into the recipient hole, allow it to protrude slightly. The protruding punches can be gently pressed into position with a clean glass slide.
25. Gently wipe away any excess paraffin from the array block after the placement of each donor punch.
26. Move across the array horizontally as the *x*-axis precision guide is closer to the operator and more convenient. To move the array from front-to-back, use the *y*-axis precision guide located at the back of the arrayer.
27. The tape-transfer method captures sections flat and uncompressed as they are cut. Cut sections are transferred to an adhesive coated slide. Sections are immediately ready for deparaffinization, eliminating the need for the water bath and drying steps. This method has been shown to cause uneven staining of TMAs in automated immunostainers.
28. To preserve TMA antigenicity combined paraffin coating and nitrogen storage can be used.

Acknowledgments

We would like to thank Elton Rexhepaj for providing the images in Fig. 2.

References

1. Battifora H. (1986) The multi-tumour (sausage) tissue block. Novel method for immunohistochemical antibody testing. Lab Invest. **55**, 244–8.
2. Kononen, J., Bubendorf, L., Kallioniemi, A., et al. (1998) Tissue microarrays for high-throughput molecular profiling of tumor specimens. Nat Med **4**, 844–7.
3. Simon, R. and Sauter, G. (2002) Tissue microarrays for miniaturized high-throughput molecular profiling of tumours. Exp Hematol **30**, 1365–1372.
4. Voduc, D., Kenney, C., and Nielsen, T. (2008) Tissue Microarrays in Clinical Oncology. Semin Radiat Oncol. **18**, 89–97.
5. Chen, W., and Foran, D. (2006) Advances in cancer tissue microarray technology: Towards improved understanding and diagnostics. Anal Chim Acta **564**, 74–81.
6. Bubendorf, L., Nocito, A., Moch, H., et al. (2001) Tissue microarray (TMA) technology: miniaturized pathology archives for high-throughput in situ studies. J Pathol. **195**, 72–9.
7. Moch, H., Kononen, T., Kallioniemi, O.P., and Sauter, G. (2001). Tissue microarrays: What will they bring to molecular and anatomic pathology? Adv. Anat. Pathol. **8**, 14–20.
8. Rimm, D.L., Camp, R.L., Charette, L.A., et al. (2001) Tissue Microarray: A new technology for amplification of tissue resources. Cancer J. 7, 24–31.
9. Packeisen, J., Korsching, E., Herbst, H., et al. (2003) Demystified Tissue microarray technology. J Clin Pathol: Mol Pathol. **56**, 198–204.
10. Jacquemier, J., Ginestier, C., Charafe-Jauffret, E., et al. (2003) Small but high throughput: how “tissue-microarrays” became a favorite tool for

pathologists and scientists. Ann Pathol. **23**, 623–32.

11. Fedor, H. L. and De Marzo, A.M. (2005) Practical Methods for Tissue Microarray Construction. Methods in Molecular Medicine. Pancreatic Cancer: Methods and Protocols. Vol. **103**, 89–101.
12. Hewitt, S. M. (2004) Design, Construction, and Use of Tissue Microarrays. Methods in Molecular Biology. Protein Arrays: Methods and Protocols Vol. **264**, 61–72.
13. Torhorst, J., Bucher, C., Kononen, J., et al. (2001) Tissue microarrays for rapid linking of molecular changes to clinical endpoints. Am J Pathol **159**, 2249–56.
14. Rubin, M., Dunn, R., Strawderman, M., and Pienta, K.J. (2002) Tissue microarray sampling strategy for prostate cancer biomarker analysis. Am. J. Surg. Pathol. **26**, 312–19.
15. Gillett, C.E., Springall, R.J., Barnes, D.M., and Hanby, A.M. (2000) Multiple tissue core arrays in histopathology research: a validation study. J. Pathol. **192**, 549–553.
16. Camp, R.L., Charette, L.A., and Rimm, D.L. (2000) Validation of tissue microarray technology in breast carcinoma. Lab Invest. **80**, 1943–1949.
17. DiVito, K.A., Charette, L.A., Rimm, D.L., and Camp, R.L. (2004) Long-term preservation of antigenicity on tissue microarrays. Lab Invest. **84**, 1071–1078.
18. Rexhepaj, E., Brennan, D.J., Holloway, P., et al. (2008) Novel image analysis approach for quantifying expression of nuclear proteins assessed by immunohistochemistry: application to measurement of oestrogen and progesterone receptor levels in breast cancer. Breast Cancer Res. **10**, R89.

Chapter 11

Immunohistochemical and Immunofluorescence Procedures for Protein Analysis

Kishore Reddy Katikireddy and Finbarr O'Sullivan

Abstract

Immunohistochemistry (IHC) and immunofluorescence (IF) involve the binding of an antibody to a cellular or tissue antigen of interest and then visualisation of the bound product by fluorescence/with the 3,3′-diaminobenzidine (DAB) chromogen detection system. With increasing numbers of available antibodies against cellular epitopes, IHC and IF are very useful diagnostic tools as well as a means to guide specific therapies that target a particular antigen on cell/tissue samples.

There are several IHC and IF staining methods that can be employed depending on the type of specimen under study, the degree of sensitivity required, and the cost considerations. The following is a basic "generic" method for localising proteins and other antigens by direct, indirect, IHC and IF. The method relies on proper fixation of tissue/cells to retain cellular distribution of antigen and to preserve cellular morphology. Details of reagents required are outlined. Consideration is also given to artefacts and other potential pitfalls and thus means to avoid them.

Key words: Immunohistochemistry, Immunofluorescence, Fixatives, Antigen, Antibody, Antigen retrieval, Reagents

1. Introduction

Immunohistochemistry (IHC) is the localisation of antigens, usually proteins, in tissue sections and cells by the use of an antibody (antibodies) with specificity for an antigen (antigens). These are subsequently visualised by a marker such as an enzyme forming a colour precipitate. Immunofluorescence (IF) is a specialised type of IHC that uses a fluorescent dye to visualise antibody binding. The localisation of a protein in tissue and cells and any variations that occur provide insights into the function of that protein. Albert H. Coons et al. (1, 2) were the first to label antibodies with a fluorescent dye and subsequently use them to identify antigens in tissue

Lorraine O'Driscoll (ed.), *Gene Expression Profiling: Methods and Protocols*, Methods in Molecular Biology, vol. 784,
DOI 10.1007/978-1-61779-289-2_11,

sections. With the expansion and development of IHC techniques, enzyme labels have been introduced. These include enzymes such as peroxidise (3, 4) and alkaline phosphatise (5). Colloidal gold (6) label has also been used to identify immunohistochemical reactions at both light and electron microscopy level. One can choose between monoclonal and polyclonal antibodies raised against either the purified or at least partially purified (7, 8) receptors or ligands or, alternatively, raised against synthetic peptides (10–20 amino acid residues long). This chapter aims to provide basic methods for IHC and IF.

1.1. Fixation

The first and most fundamental step of good IHC or IF is the selection of the optimum processing and fixation method (9, 10). Use of an inappropriate fixative may lead to misleading or artificial staining patterns. Therefore, it is advisable when starting with a new antibody to begin optimisation with a number of fixation methods. The purpose of fixation is to prevent loss stabilise structures within tissues and cells. Unfortunately, there is no perfect fixative and an inappropriately selected fixative may destroy the binding site on the antigen that the antibody recognises or degrade a structure within a cell (11, 12). The two main methods for fixation are protein precipitation using solvents and chemical cross-linking. Solvent extraction is most commonly used in IF. The two solvents most often used are methanol and acetone, both at –20°C. These are quite harsh fixatives, especially acetone. The second form of fixation, chemical cross-linking, usually is via the use of aldehydes, though other chemical compounds, for example ethylene glycol bis(succinimdylsuccinate) are occasionally used. Formaldyde-based fixation is the most popular of the aldehyde fixation methods used and works primarily by cross-linking basic amino acids with methylene bridges. Another aldehyde sometimes used is Glutaraldehyde, which is a popular fixative in electron microscopy. It is a very efficient cross linker and may be beneficial the antigen is of low molecular weight. However, it can generate a lot of background autofluorescence and is often used at low concentrations in combination with formaldehyde in IF applications.

After fixation, the tissue or cells can be embedded in paraffin and sectioned, or for frozen embedded in optimum cutter temperature (OCT)-compound. Frozen sections are most often used for immunofluorescent studies. The sectioning of tissues and cells mean a permeabilisation step is not required. However, for cultured cells stained on glass slides such a step is required to visualise intracellular antigens.

1.2. Detection

The business end of IHC is the detection of an antigen using an antibody. Other probes are also occasionally used such as lectins in the identification of glycans. The antibody classes predominantly used are IgG and IgM and these may be polyclonal or monoclonal.

Polyclonal antibodies are a mixture of immunoglobulins that are derived from many different cells. This mixture of antibodies recognise different epitopes on the same antigen, and are easier to generate then monoclonal antibodies. Monoclonal antibodies are identical immunoglobulins that are derived from a clone of a B-cell. These monoclonal antibodies react to a specific epitope on an antigen. The decision to use a polyclonal or monoclonal antibody in a study will depend on a number of factors. These include the availability of an antibody from commercial suppliers and the type of study being conducted. The main advantage of monoclonal antibodies is their homogeneity and minimal batch-to-batch variation. The homogeneity of monoclonal antibodies makes them especially useful for investigating protein–protein interactions, phosphorylation states, and distinguishing between members of protein families. In contrast the heterogeneity of polyclonal antibodies means they can have better specificity for an antigen, show more stability over a range of pH and salt concentrations.

In order to detect the binding of the antibody to its antigen either a direct or indirect method can be used. In the direct method, the primary antibody is labelled with an enzyme or in the case of fluorescence a fluorophore (Fig. 1). In the indirect method, the primary antibody is unlabelled and its binding is detected by a secondary antibody that has an enzyme or a fluorophore label (Fig. 2). The indirect method amplifies the signal as a number of secondary antibodies can bind to the primary antibody, hence increasing the amount of product deposited at the site of the antigen in the case of enzyme detection or the amount of fluorophore present. The indirect method can also be a three step procedure [known often

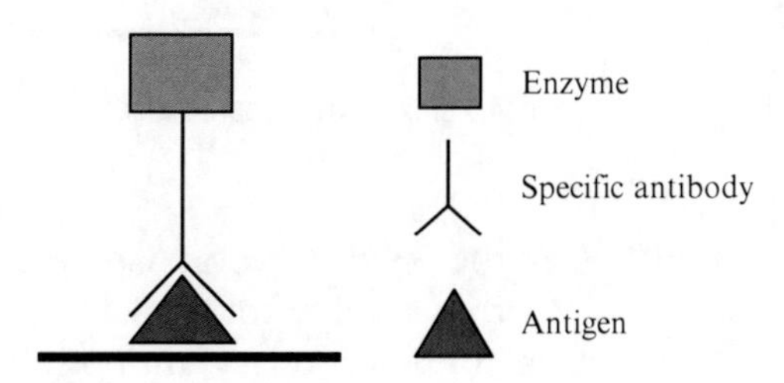

Fig. 1. Direct immunoenzyme technique.

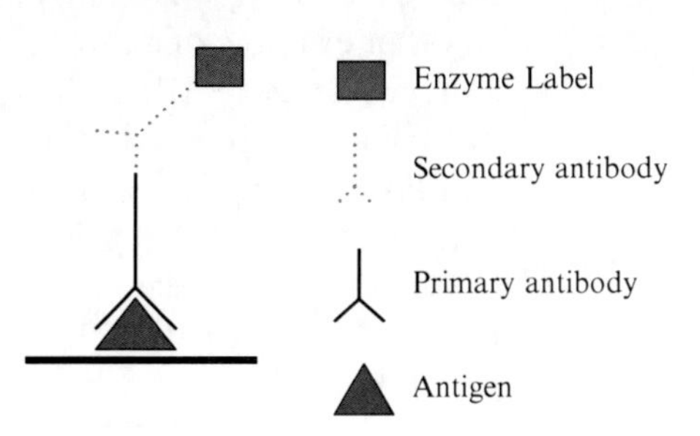

Fig. 2. Indirect immunoenzyme technique.

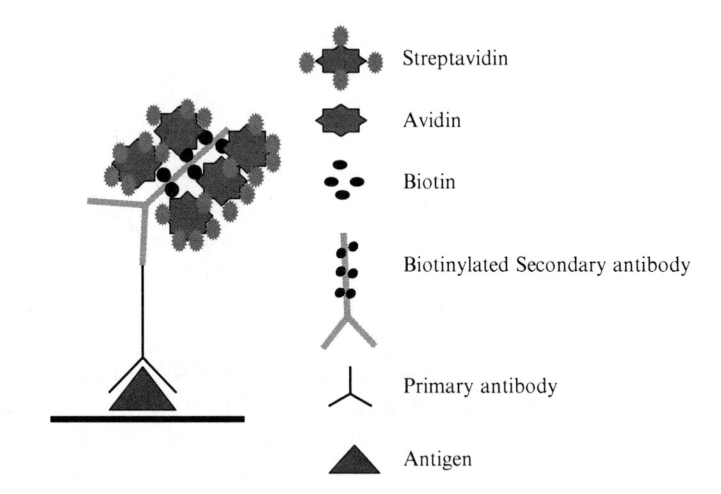

Fig. 3. Avidin–biotin complex method.

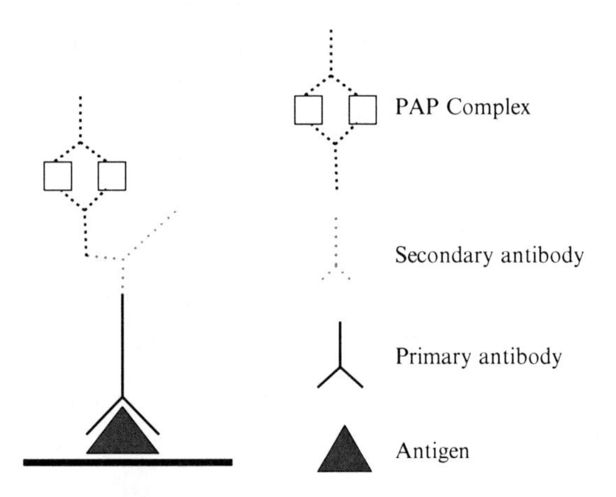

Fig. 4. Peroxidase–antiperoxidase complex.

as the Avidin–Biotin Complex (ABC) method] which helps to further amplify the signal. One popular three step method is the use of a secondary antibody which is labelled with biotin. This biotin label can bind a number of avidin/strepavidin complexes that have been linked to an enzyme or fluorophore (Fig. 3). Another example of a three step indirect method is the peroxidase–antiperoxidase (PAP) technique. The PAP method utilises a secondary antibody to form a bridge between the primary antibody and PAP complex (Fig. 4). The PAP complex consists of a combination of antibody against horseradish peroxidase and several horseradish peroxidase molecules. Using this approach, the primary antibody may be highly diluted, thereby reducing background staining. Excellent commercial sources for rabbit, goat, and mouse PAP complex are currently available. As the secondary antibody binds specifically to the Fc components of the primary and tertiary reagents, the PAP complex must be made from the same species used for preparation of the primary antibody.

There is a wide range of labels available. Enzyme-based labels work by converting a soluble form of a chromogen to an insoluble coloured precipitate. The most common label of choice is probably horseradish peroxidase; suitable chromogens include 3′,3′-diaminobenzidine (DAB), 3-ethylcarbole (AEC), and 4-chloro-1-napthol. Alkaline phosphatase is also widely used; suitable chromogens include fast red/fast blue B. Advantages of these enzyme methods include the requirement for only a normal light microscope, the precipitate (particularly DAB) remains strong for a number of years allowing samples to be archived. In IF probes are attached to an antibody or streptavidin/avidin complex that fluoresces when excited by a specific wavelength of light. Originally, fluorescein (FITC), TRITC, and rhodamine were the molecules commonly used. Today there is a extensive range of fluorophores available, which provide greater brightness and greater photostability. These include the alexa dyes from Invitrogen, the Cy dyes from GE, and quantum dots, which are fluorescent semiconductor nanoparticles, also from Invitrogen. Fluorescence requires a specialised microscope and suitable filter sets for the fluorescent probes being used. Fluorescence is method of choice when localising one or more proteins with a cell. It also allows advanced techniques such as confocal microscopy to be used. A disadvantage, however is the signals will at most only last for about 10 days.

There are numerous IHC methods that may be used to localise antigens. The selection of a suitable method should be based on parameters such as the type of specimen under investigation and the degree of sensitivity required.

2. Materials

2.1. Fixatives

1. 4% Paraformaldehyde (Sigma): add 40 g paraformaldehyde to 500 mL distilled water and heat slowly to 70°C. Add 2 M NaOH (dropwise) until solution is clear, allow to cool and add 0.2 M phosphate buffer. Adjust pH to 7.4 and filter. Ideally, prepare fresh; however, aliquots may be stored at −20°C for a short (less than a week) period of time. Do not re-freeze (see Note 1).
2. Methanol (Merck) at −20°C for 5 min followed by acetone at −20°C for 30 s. Alternatively, acetone (Sigma) alone at −20°C for 30 s may be suitable (see Note 2).
3. Mounting medium: Fluormount G (Southern Biotechnology) or prepare 50% (w/v) glycerol (Sigma) and 0.1% (w/v) *p*-phenylenediamine (Sigma) in phosphate-buffered saline (PBS), pH 8.0.

2.2. Buffers

1. Sodium citrate buffer: 10 mM sodium citrate, 0.05% Tween 20, pH 6.0.

   Tri-sodiumcitrate(dihydrate) (Sigma) – 2.94 g.

   Distilled water 1,000 mL.

   Mix to dissolve. Adjust pH to 6.0 with 1N HCl.

   Add 0.5 mL of Tween 20 (Sigma) and mix well. Store at room temperature for 3 months or at 4°C for longer storage.

2. 1 mM EDTA, adjusted to pH 8.0.

   EDTA (Sigma) - 0.37 g.

   Distilled water 1,000 mL.

   Store at room temperature for 3 months.

3. Tris–EDTA Buffer: 10 mM Tris Base, 1 mM EDTA solution, 0.05% Tween 20, pH 9.0.

   Tris – 1.21 g.

   EDTA – 0.37 g.

   Distilled water 1,000 mL (100 mL to make 10×, 50 mL to make 20×).

   Mix to dissolve. pH is usually at 9.0.

   Add 0.5 mL of Tween 20 and mix well. Store at room temperature for 3 months or at 4°C for longer storage.

4. PBS: 0.1 M NaH_2PO_4 (Sigma), 0.1 M Na_2HPO_4 (Sigma), and 0.15 M NaCl (Sigma) in distilled water, pH 7.4 (see Note 3).

5. PHEM Buffer [PIPES, HEPES, EGTA, $MgCl_2$ (Sigma, St. Louis, MO)]: To prepare a stock solution, dissolve three pellets of NaOH in 75 mL water and add 3.63 g PIPES. Bring to pH 7. Then add 1.19 g HEPES, 0.08 g $MgCl_2 \cdot 6H_2O$, and 0.76 g EGTA and bring pH to 6.9 with 1N NaOH. Bring final volume to 100 mL. Dilute with distilled water in 1:1 ratio to 0.1 M before use.

6. Buffered glycerol: adjust the pH of 0.5 M Na_2CO_3 solution to 8.6 with 0.5 M $NaHCO_3$. Add one part of this buffer to two parts 100% glycerol. Store at 4°C for up to 2 months.

2.3. Equipments

1. Domestic (850 W) or scientific microwave.
2. Microwaveable vessel with slide rack to hold approximately 400–500 mL or Coplin jar.
3. Hot plate.
4. Light microscope and fluorescence microscope (with 4, 10, 20, and 40× objective lenses, as relevant for the study undertaken).

3. Methods

Any reagent stored at 4°C must be allowed to equilibrate to room temperature before use.

3.1. Immunohistochemistry on Paraffin Sections

3.1.1. Deparaffinisation and Rehydration of Tissue Sections, Blocking of Endogenous Peroxidase Activity

Before proceeding with the staining protocol, tissue sections from fixed and paraffin-embedded material must first be deparaffinised and rehydrated. Incomplete removal of paraffin can cause poor staining of the section.

1. Circle and label the specimen with a diamond pencil.
2. Place in 60°C oven for 30 min.
3. Transfer immediately to a fresh xylene bath for 3 min.
4. Repeat step 3 above with a second xylene bath (see Note 4).
5. Place in a fresh bath of absolute alcohol for 3 min.
6. Repeat step 5 above with a second bath of absolute alcohol.
7. Place in a bath with 95% ethanol for 3 min.
8. Repeat step 7 with a second 95% ethanol bath.
9. Rinse under gently running water.
10. Transfer into PBS.

3.1.2. Antigen Retrieval by Microwave Method

1. Deparaffinise and rehydrate the sections as above.
2. Add the appropriate antigen retrieval buffer such as sodium citrate buffer (pH 6.0) to the microwaveable vessel (see Note 5).
3. Remove the slides from the tap water and place them in the microwaveable vessel. Place the vessel inside the microwave. If using a domestic microwave, set to full power and wait until the solution boils. Boil for 20 min from this point. If using a scientific microwave, program so that antigens are retrieved for 20 min once the temperature has reached 98°C (see Note 6).
4. When 20 min has elapsed, remove the vessel and run cold tap water into it for 10 min. Use hot solution with care (see Note 7).

3.1.3. Staining Using Indirect Method

All incubations should be carried out in a humidified chamber to avoid drying of the tissue. This procedure can be used for cells cultured on coverslips or slides. However, they must be appropriately fixed and permeabilised prior to starting (see steps 1–5 of Subheading 3.2).

1. Wash the slides (2 × 5 min) in PHEM/PBS containing 0.025% Triton X-100, with gentle agitation (see Note 8).

2. Wipe off excess buffer from slide and, using a PAP pen, draw a circle around the tissue (it prevents the waste of valuable antibody by creating a water repellent circle around the section).
3. Apply 10% normal serum (NS) solution that corresponds to the antibody being used on that slide (i.e. mouse serum in association with mouse primary antibodies).
4. Incubate at room temperature for 10–20 min.
5. Gently tap off excess serum. Do not rinse.
6. Apply enough primary antibody (primary antibody diluted in 1% NS or 1% BSA) to cover the tissue (see Note 9).
7. Incubate overnight at 4°C for the length of time appropriate for the particular antibody (based on manufacturer's recommendations and in-house optimisation) (see Note 10).
8. Rinse in PHEM/PBS 0.025% Triton (2 × 5 min), with gentle agitation.
9. If using an HRP conjugate for detection, incubate the slides in 0.3% H_2O_2 in PHEM/PBS for 15 min (this is a blocking step to avoid non-specificity of HRP conjugate).
10. For enzymatic detection (fluorophore secondary conjugates), apply enzyme-conjugated secondary antibody to the slide (diluted to the concentration recommended by the manufacturer) in PHEM/PBS with 1% wt/vol BSA, and incubate for 1 h at room temperature.
11. Rinse 3 × 5 min PHEM/PBS (see Note 11).
12. Develop the coloured product of the enzyme with the DAB chromogen. To prepare working solution: transfer 2 mL DAB stock solution to another test tube, and subsequently add 15 μL of 3% hydrogen peroxide. Once prepared, working solution is only stable for 2 h (see Note 12).
13. Rinse in running tap water for 5 min.
14. Counterstain (if required) (see Note 13).

3.1.4. Dehydrating and Cover-Slipping

1. Dip slides in each reservoir of the reagent container tray; beginning with distilled water and working backwards through 95% ethanol, absolute ethanol, and then xylene. Keep slides in xylene until ready to coverslip.
2. Place a small amount of DPX-mounting media onto the tissue section using a plastic transfer pipette.
3. Drop the appropriate size coverslip onto the section and press down to allow any trapped air bubbles to float to outer edges of cover slip.

4. Turn slide over onto a 4×4 gauze pad and press to absorb excess DPX mount.
5. Dip briefly in xylene to remove any excess DPX-mounting meida or fingerprints. Air dry.

3.2. Immuno-fluorescence of Cultured Cells

This is only an outline of the procedure for IF; various steps will require optimisation for a particular antibody. All steps are carried out in a moist chamber.

1. Cells are allowed to adhere to glass coverslips overnight (see Note 14).
2. Rinse the coverslips 3×5 min in PHEM buffer pH 7–8.
3. Fix the cells with the fixative of choice (e.g. 20 min in 4% para-formaldehyde and 0.5% glutaraldehyde at room temperature or 5 min ice cold methanol at −20°C).
4. Wash the coverslips 3×5 min with PHEM buffer.
5. Cells are permeabilised with either 0.1% Triton X-100, 0.5% Triton X-100, methanol or acetone (see Note 15).
6. Quench autofluorescence if necessary with 0.1% (w/v) sodium borohydride in PHEM (ice cold). Apply while fizzing 3×10 min at ice/room temperature. Warning explosive hydrogen liberated when wet (see Note 16).
7. Slides are then rinsed 2×10 min with PHEM buffer.
8. Non-specific binding sites are blocked by incubating with 5% (v/v) NS in PHEM buffer for 30 min at room temperature A detergent (e.g. 0.025% Tween 20) should be included, if the anti-serum being used was raised against an intracellular antigen or domain (see Note 17).
9. Remove the blocking sera from the cells and incubate with the primary antibody diluted in 1% (v/v) NS PHEM for the optimum time e.g. overnight at 4°C *or* room temperature for 2 h *or* 37°C for 1 h (see Note 18).
10. Wash the coverslips 3×5 min in PHEM buffer at room temperature.
11. Incubate coverslips with secondary antibody diluted in 1% v/v NS PHEM for 1 h at room temperature in darkness.
12. If required, counterstain Nuclei with DAPI solution (1:20,000 dilution) for 5 min.
13. Wash coverslips 3×5 min in PHEM buffer.
14. Mount the coverslip on with mounting media such as Pro-Long Gold or Vector shield. Seal with clear nail varnish. Examine the section with an epifluorescence microscope equipped with the appropriate filter for the chosen dye.

3.3. Controls

1. Controls must be run in order to test the protocol and for the specificity of the antibody being used.
2. Positive controls should be included to test a protocol or procedure and make sure it works. It is thus ideal to use the tissue of known positive as a control. If the positive control tissue showed negative staining, the protocol or procedure needs to be checked and good positive staining is reproducibly obtained.
3. Negative controls should be included to test for the specificity of an antibody involved. Firstly, no staining must be shown when omitting primary antibody or replacing a specific primary antibody with NS (must be the same species as primary antibody). This control is easy to achieve and can be used routinely in immunohistochemical staining. Secondly, the staining must be inhibited by adsorption of a primary antibody with the purified antigen prior to its use, but not by adsorption with other related or unrelated antigens. This type of negative control is ideal and necessary in the characterisation and evaluation of new antibodies, but it is sometimes difficult to obtain the purified antigen; therefore, it is rarely used in immunohistochemical staining. Tissue or cells which do not express test antigen might be another control (optional).
4. When performing double labelling in IF, it is important to add the following controls. Primary A incubated with the secondary for Primary B and vice versa. These controls for crosstalk from one primary to the other secondary, the sample should be unstained. Primary A incubated with the Secondary for primary A and Primary B, with same incubation for Primary B. This control for the secondary's binding to each other the sample should only stain for one colour.

4. Notes

1. PFA should always be prepared fresh on the same day you wish to use it. Storage overnight at 4°C is possible, but it will not fix as well the second day. It is possible to freeze the PFA solution at –20°C, but for consistency and reproducibility of results, tissue should either be fixed always with fresh PFA or always with freshly thawed PFA. Of importance, formalin fixation is good for IHC, but not advised for IF as it generates high autofluorescence background.
2. When selecting fixatives, due considerations must be given to the nature and specifics of the antigen of interest. For example, methanol is good for maintaining actin and other cytoskeleton proteins. Conversely, it may destroy glycolipid antigens by leaching them out of the cells.

3. The solution should be at pH 7.4. Do not pH using acid or base. If you need to adjust the pH, make up a separate 0.2 M solution of either the monobasic or dibasic sodium phosphate (depending on how you need to adjust the pH) and add accordingly.
4. Xylene is carcinogenic, DO NOT inhale. Keep the slides in the tap water until ready to perform antigen retrieval. At no time from this point onwards the slides should not be allowed to dry. Drying out will cause non-specific antibody binding, and therefore high background staining.
5. Use a sufficient volume of antigen retrieval solution in order to cover the slides by at least a few centimetres if using a non-sealed vessel to allow for evaporation during the boil. Be sure to watch for evaporation and for boiling over during the procedure, and do not allow the slides to dry out. With this method, antigens can be significantly improved by the pre-treatment with the antigen retrieval reagent that break the protein cross-links formed by formalin fixation, and thereby uncover hidden antigenic sites.
6. 20 min is only a suggested antigen retrieval time. Less than 20 min may leave the antigens un-retrieved, leading to weak staining. More than 20 min may leave them over-retrieved, leading to non-specific background staining and also increasing the chances of sections dissociating from the slides. A control experiment is recommended beforehand, where slides containing consecutive sections from the same tissue are retrieved for 5, 10, 15, 20, 25, and 30 min before being stained, by IHC. This should enable identification of optimum antigen retrieval time for the particular antibody being used.
7. This allows the slides to cool enough so that they can be handled, and allows the antigenic site to re-form after being exposed to high temperature.
8. The use of 0.025% Triton X-100 in the TBS helps to reduce surface tension, allowing reagents to cover the whole tissue section with ease. It is also believed to dissolve Fc receptors, therefore reducing non-specific binding.
9. The dilution of the antibody varies for each individual antiserum. In general, purified IgGs are used at concentrations ~1 μg/mL, but it may be a factor 10 up or down. Even more difficult to predict is the dilution at which antisera are used. One has to compare different dilutions for specific and non-specific signal and identify the optimum.
10. Overnight incubations allow antibodies of lower titre or affinity to be used by simply allowing more time for the antibodies to bind. Also, regardless of the antibody's titre or affinity for its target, once the tissue has reached saturation point no more

binding can take place. Overnight incubation ensures that this occurs.

11. If using fluorescence detection, the procedure ends at this step and a coverslip is applied with mounting medium. If visualising the protein with a chromogen, continue with the following steps.
12. DAB is a suspected carcinogen. Dispose it according to good laboratory guidelines and local regulatory guidelines.
13. Commonly used counterstains include haematoxylin (blue); nuclear fast red; and methyl green. When using fluorescence detection, DAPI (blue) or propidium iodide/PI (red) can be used to counterstain in immunofluorescence.
14. Use a 1.5 glass coverslip as most microscope objectives are corrected for this thickness, hence this will allow optimum imaging. Prior to use, wash the coverslips in a weak acid solution (1 N HCl) and sterilise using an oven (120°C for 2 h). Alternatively, soak the coverslip in 95% ethanol and gently flame them. This removes oils that coat the coverslip and helps promote cell adherence. The use of round coverslips in 24-well plates allows convenient culture of the cells.
15. It is necessary to permeabilse cells that have been aldehyde fixed. A number of permeabilisation options exist, and it is advised that it should be optimised according to assay. Methanol- or acetone-fixed material is already permeabilised, but such material may still benefit from permeabilisation solutions in wash buffer.
16. Aldehyde fixation often results in increased autofluorescence sodium borohydride (0.1% (w/v)) is one means to quench this. An alternative method is to use 1% (w/v) or 150 mM glycine with 0.1% Tween in PBS or PHEM buffer.
17. The NS used should be from the same species as the secondary antibody.
18. Incubation time is determined by the antibody specificity and concentration. It is worth spending time in optimising this step. If double labelling is performed, sometimes better staining patterns are observed when the primary antibodies are added sequentially. Similarly for secondary antibodies.

References

1. Coons, A.H., Leduc, E.H., and Connolly, J.M. (1955) Studies on antibody production. I. A method for the histochemical demonstration of specific antibody and its application to a study of the hyperimmune rabbit. *J Exp Med.* **102,** 49–60.
2. Kaplan, M.E., Coons, A.H., and Deane, H.W. (1950) Localization of antigen in tissue cells;

cellular distribution of pneumococcal polysaccharides types II and III in the mouse. *J Exp Med.* **91**, 15–30.

3. Nakane, P.K., and Pierce, G.B. (1966) Jr. Enzyme-labeled antibodies: preparation and application for the localization of antigens. *J Histochem Cytochem.* **14**, 929–31.
4. Avrameas, S., and Uriel, J. [Method of antigen and antibody labelling with enzymes and its immunodiffusion application]. (1966) *C R Acad Sci Hebd Seances Acad Sci D.* **262**, 2543–2545.
5. Mason, D.Y., and Sammons, R. (1978) Alkaline phosphatase and peroxidase for double immunoenzymatic labelling of cellular constituents. *J Clin Pathol.* **31**, 454–60.
6. Faulk, W.P., and Taylor, G.M. (1971) An immunocolloid method for the electron microscope. Immunochemistry. **8**, 1081–3.
7. Maderspach, K., Nemeth K., Simon, J., Benyhe, S., Szucs, M., and Wollemann, M.A. (1991) Monoclonal antibody recognizing kappa- but not mu- and delta-opioid receptors. J Neurochem. **56**, 1897–904.
8. Pichon, J., Hirn, M., Muller, J.M., Mangeat, P., Marvaldi, J. (1983) Anti-cell surface monoclonal antibodies which antagonize the action of VIP in a human adenocarcinoma cell line (HT 29 cells). *Embo J.* **2**, 1017–22.
9. Somogyi, P., Smith, A.D., Nunzi, M.G., Gorio, A., Takagi, H., and Wu, J.Y. (1983) Glutamate decarboxylase immunoreactivity in the hippocampus of the cat: distribution of immunoreactive synaptic terminals with special reference to the axon initial segment of pyramidal neurons. *J Neurosci.* **3**, 1450–68.
10. Stefanini, M., De Martino, C., and Zamboni, L. (1967) Fixation of ejaculated spermatozoa for electron microscopy. *Nature.* **216**, 173–4.
11. Melan, M., and Sluder, G. (1992) Redistribution and differential extraction of soluble protein in permeabilized cultured cells. *J Cell Sci.* **101**, 731–743.
12. Hannah, M., Weiss, U. and Huttner, W. (1998) Differential extraction of protein from paraformaldehyde fixed cells: Lessons from synaptophysin and other membrane proteins, Methods. **16**, 170–181.

Chapter 12

Advanced Microscopy: Laser Scanning Confocal Microscopy

Orla Hanrahan, James Harris, and Chris Egan

Abstract

Fluorescence microscopy is an important and fundamental tool for biomedical research. Optical microscopy is almost non-invasive and allows highly spatially resolved images of organisms, cells, macromolecular complexes, and biomolecules to be obtained. Generally speaking, the architecture of the observed structures is not significantly modified and the environmental conditions can be kept very close to physiological reality. The development of fluorescence microscopy was revolutionized with the invention of laser scanning confocal microscopy (LSCM). With its unique three-dimensional representation and analysis capabilities, this technology gives us a more real view of the world.

This chapter introduces the reader to the methodology of setting up basic experiments for use with a laser scanning confocal microscope. There are practical guidelines about sample preparation for both fixed and living specimens, as well as examples of some of the applications of confocal microscopy.

Key words: Confocal microscopy, Fluorescence microscopy, Imaging, Organelle markers

1. Introduction

Laser scanning confocal microscopy (LSCM) has become an invaluable tool for a wide range of investigations in the biological and biomedical sciences. There are a number of advantages to using confocal microscopy over conventional wide-field microscopy, including the elimination or reduction of out-of-focus light (that leads to image degradation), the ability to control depth of field, and the capability to collect serial optical sections from thick specimens. Essentially, confocal microscopy can reveal the three-dimensional structure of a specimen.

In conventional wide-field fluorescence microscopy, the entire depth of the fluorescent specimen is illuminated. The fluorophores throughout the thickness of the specimen are excited, and therefore fluorescence signals are collected not only from the plane of

Lorraine O'Driscoll (ed.), *Gene Expression Profiling: Methods and Protocols*, Methods in Molecular Biology, vol. 784,
DOI 10.1007/978-1-61779-289-2_12,

focus, but also areas above and below. This produces images that are often blurry and lack contrast and this is further compounded when thicker specimens are imaged (generally, anything thicker than 2 μm). A confocal microscope selectively collects light from a thin (<1 μm) optical section at the plane of focus in the specimen. This is accomplished by scanning the specimen with a focussed beam of light and collecting the fluorescence signals emitted by the specimen via a pinhole aperture. Structures within the focal plane appear more sharply defined because there is no glare of light from out-of-focus areas.

Confocal imaging is feasible in both living and fixed specimens. It is possible to study dynamic processes, such as gene expression (1), cytoskeletal assembly and turnover, chromosome dynamics, and molecular binding interactions (2). In addition, by using a number of different fluorescent markers for various cell organelles, it is possible to localize a protein of interest to a particular structure within the cell (3, 4).

Specimen preparation is critical for obtaining good images. Unfortunately, some people assume that a confocal microscope can make poorly stained specimens generate informative data. Confocal microscopy makes images sharper by rejecting out-of-focus light, but it cannot discriminate between specific and non-specific staining or between a specific probe and autofluorescent background. It takes a picture of all the light coming from a selected focal plane. If anything, a confocal will render artefacts sharper and more beautiful. Therefore, good staining protocols and adequate controls are more important than ever. For a comprehensive reference book on confocal microscopy, read "The Handbook of Biological Confocal Microscopy", *edited by James B. Pawley. Plenum Press, New York, 1995.* This is an extensive, well written, overview of the history, design, applications, and reference to the use of confocal microscopy in all its forms in biological research and education.

2. Materials

2.1. Cell Culture (Adherent and Suspension Cells)

1. Tissue culture flasks.
2. Appropriate cell culture media.
3. Chamber coverglass slides (Lab-Tek chambered coverglass w/cover # 1.5 borosilicate sterile, NUNC).
4. Glass bottom dishes (WillCo, 35/50 mm depending on the stage set up of the microscope).
5. Coverslips (22 × 22 mm, thickness no. 1.0–1.5 or 0.13–0.19 mm, available from a number of suppliers).
6. Sterile chamber slides (BD Biosciences, Cat. No. 354632; Poly-D-Lysine 8-well Culture Slides).

7. Poly-L-Lysine available from Sigma, P4707 (sterile solution) and P8920, 0.01% solution (see Notes 1 and 2).
8. Microscope slides.
9. Welled slides (5 wells per slide, Thermo Scientific).

2.2. Fixation/Permeabilization (see Note 3 for Additional Information)

1. 6% paraformaldehyde solution – *Caution*. Must be prepared in a fume hood. 6% paraformaldehyde (w/v) is prepared freshly on the day of the experiment or frozen in aliquots at −20°C. The required weight of paraformaldehyde (6 g if making 100 mL solution) is added to 50 mL of distilled, deionised water. The insoluble paraformaldehyde is then titrated with 5 M NaOH and stirred vigorously until the paraformaldehyde is fully dissolved. 40 mL PBS is then added to the fixative. 12 M HCl is added to the fixative until the pH of the solution reaches 7.5. Sufficient distilled deionised water is added to bring the solution to the required concentration.
2. PBS buffer: 136 mM NaCl, 3 mM KCl, 16 mM Na_2HPO_4, 3 mM KH_2PO_4 containing either 10 mM glucose, and 40 mM sucrose prior to fixation or 15 mM sodium azide post-fixation.
3. 100% methanol.
4. 100% acetone.
5. 1:1 methanol:acetone.
6. PBS buffer containing 0.1 M glycine or 0.1 M methylamine.

2.3. Blocking Buffer, Antibody Incubations, Organelle Marker Probes, and Mounting Media

1. Blocking buffer: PBS buffer (as above, including 15 mM sodium azide) containing between 1–5% (w/v) bovine serum albumin (BSA) and 0.5% goat serum (or serum from other species in which secondary antibody was raised).
2. For permeabilization of cells, 0.1–0.5% Triton-X-100 (depending on cell type) can be added to the blocking buffer and the primary and secondary antibody incubations. Alternatively, 0.5% saponin can be used.
3. Alexa-fluor conjugated secondary antibodies (Invitrogen).
4. Various organelle markers available from Invitrogen, e.g. MitoTracker Dye (Em λmax from 516 to 665 nm), CellMask plasma membrane stain, available as CellMask Orange and Deep Red (Ex λmax 554 nm/Em λmax 567 nm), and Hoechst/DAPI (Ex λmax 350/Em λmax 461 nm).
5. Aqueous mounting medium (see Note 4) or polymerizing mounting medium (e.g. ProLong Gold with DAPI; Invitrogen).
6. Clear nail polish.

3. Methods

The purpose of this section is to provide detailed information on how to set up experiments for confocal microscopy for both fixed and living cells and thick tissue sections.

3.1. Preparation of Adherent Mammalian Cells for Laser Scanning Confocal Microscopy

1. All solutions and equipment coming into contact with living cells must be sterile, and proper aseptic technique should be used accordingly. All culture incubations are performed in a humidified 37°C, 5% CO_2 incubator unless otherwise specified.
2. Macrophages (e.g. human peripheral blood monocyte-derived macrophages) are grown in X-Vivo 10 medium, supplemented with 2–5% autologous human serum for 12 days, IL-4 can be added to the media to induce cell–cell fusion, and formation of multinucleated giant cells (5).
3. Murine bone marrow-derived dendritic cells (BMDC) are grown from bone marrow of C57/B16 mice in RPMI with 10% FCS and 20 ng/mL GM-CSF for 11 days. The cells are fed with extra media on day 3, adherent cells isolated on day 6 and re-suspended in media with GM-CSF and fed on day 8. On day 10, non-adherent and semi-adherent dendritic cells are removed with gentle washing, re-suspended at 6.25×10^5 cells/mL in media with 10 ng/mL GM-CSF, transferred to coverslips or chamber slides and cultured for a further 24 h.
4. Bovine dendritic cells are cultured from isolated bovine peripheral blood monocytes (isolated by positive selection – using CD14 magnetic beads). The monocytes are grown in RPMI-1640 with 200 U/mL recombinant bovine IL-4 and 0.2 U/mL recombinant GM-CSF for 7 days (half of the media is replaced on day 3).
5. The cells are transferred to appropriate tissue culture vessels when they have reached the optimal cell density, typically $0.1–0.5 \times 10^6$ cells/mL (e.g. sterile poly-L-lysine coverslips (19×19 mm) in 12-well tissue culture plates for immunofluorescence studies (see Note 5). Some cells (e.g. human peripheral blood monocyte-derived macrophages) can be grown on coverslips from day 0 while others need to be transferred to coverslips at the end of the culture period.
6. Incubate cells for 1–24 h (see Note 6).
7. Remove the supernatant and non-adherent/dead cells and wash the cell layer twice with PBS.
8. Fix cells with 2–4% paraformaldehyde for 20 min at room temperature. At this point, tissue culture plates containing

coverslips for microscopy can be removed from the incubator and all further incubations can be done on the bench.

9. Wash cells three times with PBS.
10. Cells are incubated with Wheat Germ Agglutinin-Alexa 594 (5 μg/mL) for 30 min at room temperature (see Note 7).
11. Permeabilize cells by incubating with 0.1% Triton-X-100 for 5–10 min.
12. Block cells for 1 h in blocking buffer. The serum added should match the species, in which the secondary antibody was raised (i.e. use goat serum for goat anti-mouse/rabbit secondary antibodies).
13. Wash cells three times with PBS and incubate with primary antibodies for 1 h in blocking buffer at room temperature, e.g. rabbit anti-calreticulin, mouse anti-HLA-A,B,C, and mouse anti-CPVL. A number of primary antibodies can be included in this step, ensuring that all antibodies originate from a different species – e.g. a rabbit IgG and a mouse IgG (see Note 8). Alternatively, fluorescently labelled primary antibodies can be used.
14. Wash cells five times with PBS, then incubate with appropriate Alexa Fluor-conjugated secondary antibody diluted 1:2,000 in blocking buffer for 1 h at room temperature (see Note 9).
15. For aqueous mounting medium: Wash cells five times with PBS and carefully mount the specimen on a microscope slide in aqueous mounting medium containing DAPI or Hoechst to stain the nucleus (see Note 10). Seal the edge of the coverslip with clear nail polish to prevent drying.
16. For solid (polymerizing) mounting medium: Wash cells five times with PBS and carefully mount the coverslip on a drop of mounting medium, e.g. ProLong Gold with DAPI on a microscope slide and leave at room temperature overnight to set.
17. Image using laser scanning confocal microscope. An example result of such an experiment is shown in Fig. 1.

3.2. Preparation of Suspension Cells for Laser Scanning Confocal Microscopy

1. Grow cells in appropriate media to the optimal cell density. It is important to be aware that a lot of cell loss can occur due to subsequent washes and incubations, and therefore the higher the stock concentration of cells the better.
2. Optional, cell markers can be added at this point; e.g. MitoTracker Dye.
3. Wash cells in PBS buffer (3×) and fix in 2–4% paraformaldehyde solution at room temperature for 20 min (fixation conditions vary for every cell type so these should be titrated, also see Note 3).

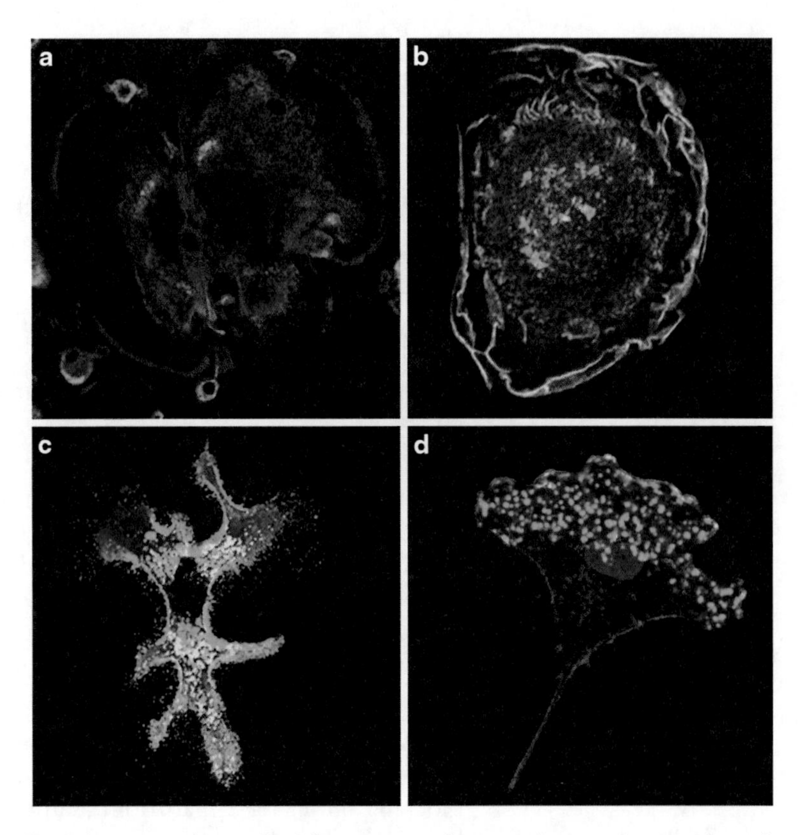

Fig. 1. Laser scanning confocal fluorescence microscopy images of mammalian cells (single optical sections). (**a**) A multinucleated giant cell, formed by the fusion of human peripheral blood monocyte-derived macrophages (MDM) treated with IL-4 (see Subheading 3). The cells are stained with TRITC-phalloidin for actin (*red*), rabbit anti-calreticulin for the ER (*green*), and mouse anti-CPVL (carboxypeptidase vitellogenic-like) (*blue*). CPVL is a serine carboxypeptidase found in the ER and membrane ruffles of macrophages (8). (**b**) Human peripheral blood MDM stained with TRITC-phalloidin (*red*) and mouse anti-human HLA-A,B,C (*green*). The picture was taken specifically to highlight the actin-rich membrane ruffles (which are very pronounced in primary human macrophages). (**c**) A murine bone marrow-derived dendritic cell (BMDC)-labelled with wheat germ agglutinin-Alexa 594 (*red*) that has phagocytosed green fluorescently labelled polystyrene micro-particles. (**d**) A bovine dendritic cell stained with FITC-phalloidin (*green*) and DAPI to identify the nuclei (*blue*) and a murine monoclonal antibody (CC20) against CD1w2 (*red*). All images display a single optical section (~3 μm) captured with a 60× oil immersion objective with a numerical aperture of 1.42. All four images display a merge of all the colour channels.

4. Wash cells with PBS (3×) and resuspend in PBS containing sodium azide, 15 mM. Fixed suspension cells can be stored at 4°C for up to 1 month without any deleterious effects on the cells morphology.
5. For immediate use, transfer fixed cells onto poly-L-lysine coverslips or poly-L-lysine coated welled slides and allow them to attach for 15 min at room temperature.

6. Remove unattached cells by dipping the slide or coverslip very gently into a solution of PBS containing sodium azide.
7. Block cells with blocking buffer for 1 h at room temperature.
8. Incubate cells with primary antibody in blocking buffer. The concentration of antibody and incubation times for all antibodies varies and this should be titrated accordingly.
9. Incubate cells with secondary antibody in blocking buffer for 1 h at room temperature. Generally, a 1 in 2,000 dilution works well for Alexa Fluor-conjugated anti-IgGs.
10. Cells are washed gently again by dipping the coverslip or slide into a solution of PBS buffer containing sodium azide.
11. Carefully mount the coverslip onto a slide with ProLong Gold mounting media containing DAPI. Leave at room temperature overnight to set.
12. Image using laser scanning confocal microscope.

3.3. Preparation of Thick Tissue Slices (e.g. Mouse Brain Tissue) for Laser Scanning Confocal Microscopy

1. Perfuse mouse with 4% paraformaldehyde (see Note 11) and extract the brain and immersion fix overnight in 4% paraformaldehyde.
2. Cut the brain tissue (15 μm thick sections) using a cyrostat and collect the sections on poly-L-lysine slides (see Note 12).
3. Permeabilize tissue sections for 30–60 min with 0.5% Triton-X-100 in PBS buffer. These conditions are specific for this particular tissue sample and need to be optimized for each new experiment.
4. Block tissue sections for 1 h at room temperature in blocking buffer. In the example here, the following blocking buffer is used: PBS containing 10% normal goat serum and 5% bovine serum albumin. Again for each tissue sample the blocking conditions need to be optimized.
5. Incubate sections overnight with rabbit polyclonal anti-GFAP IgG and mouse monoclonal anti-NeuN IgG both at a dilution of 1:200 in blocking buffer without BSA.
6. Wash tissue sections four times in PBS and incubate with Cy3-conjugated anti-rabbit IgG and Cy2-conjugated anti-mouse IgG for 2–3 h at room temperature in PBS with no BSA.
7. Wash tissue sections four times with PBS and carefully mount a coverslip onto the slide with ProLong Gold mounting media containing DAPI. Leave at room temperature overnight to set.
8. Image using laser scanning confocal microscope. An example result of such an experiment is shown in Fig. 2a, b.

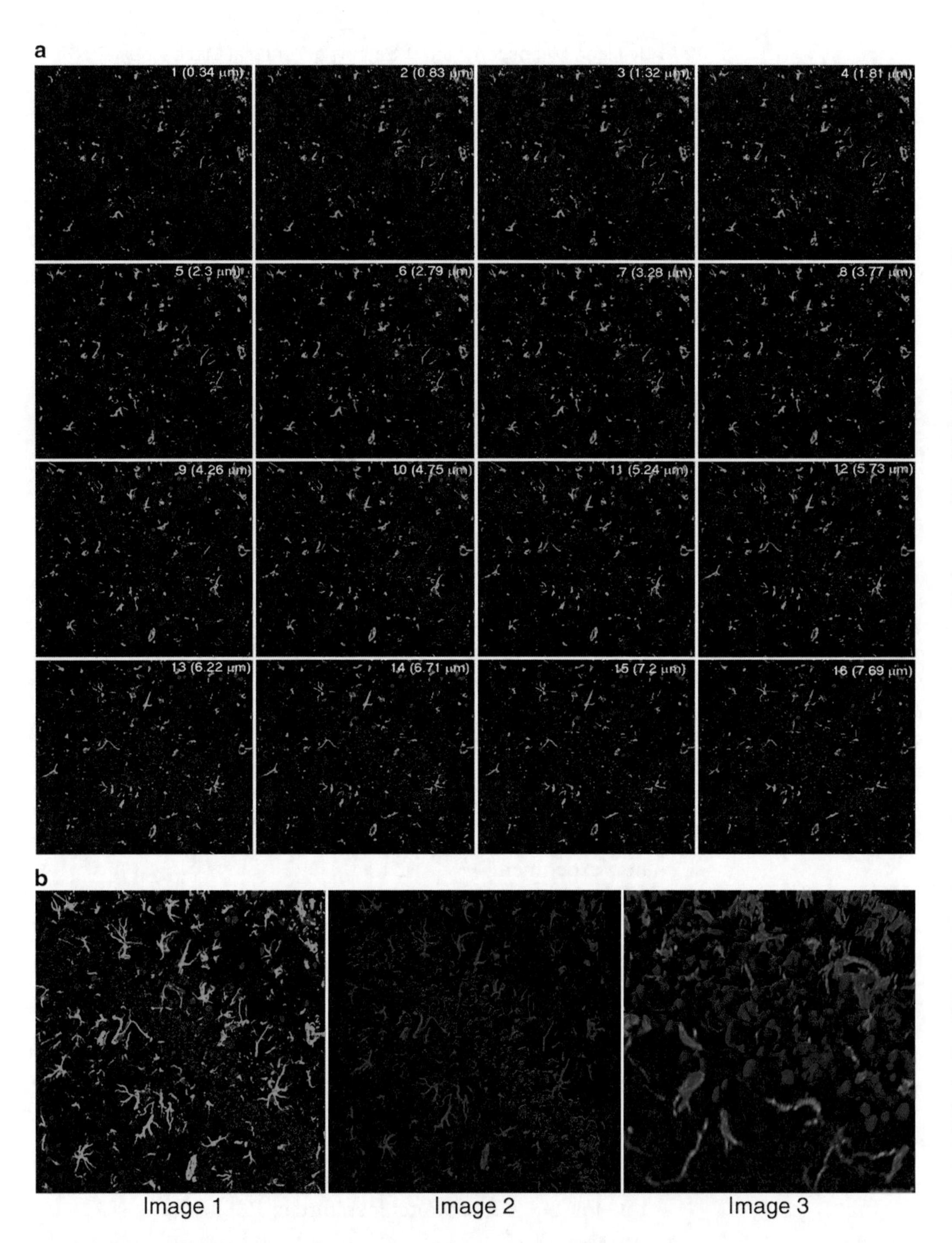

Fig. 2. Laser scanning confocal microscopy images of thick tissue specimens (optical sectioning and max projections). (**a**) Optical sections of CA1 region of adult mouse hippocampus. NeuN (*red*) is specific to neuronal cell types while GFAP (*green*) is specific predominantly to astrocytes. The section is counterstained with DAPI to highlight the nuclei (*blue*). The tissue section (15 μm) was imaged with a 40× oil immersion objective in 0.49 μm *Z*-steps. Each image in the sequence represents a single *Z*-section (0.49 μm thick) from the optically sliced tissue sample. (**b**) Image 1 – A maximum intensity projection (MIP) of the 16 optical sections in (**a**) collected at 0.49-μm intervals in the *z*-axis. Image 2 – The MIP image was iso-surface rendered using the image analysis software IMARIS. This shows with more clarity where the NeuN is located in the tissue section. Image 3 – The same picture as in image 2 which has been zoomed to show the three-dimensional effect following reconstruction of the *Z*-sections after optical sectioning on the confocal microscope and iso-surface rendering using IMARIS.

3.4. Preparation of Adherent Cells for Live Cell Imaging Using Laser Scanning Confocal Microscopy

1. Grow cells on 35-mm glass bottom Petri dishes or Lab-Tek chamber slides. When cells have reached the desired confluence for imaging (see Note 5), the dishes are removed from the incubator and placed on a heated microscope stage and the microscope is set up for live imaging (see Note 13).
2. Optional: Prior to taking the cells to the microscope for live imaging, specific cell organelle markers can be added in order to identify different structures within the cells during the imaging process, e.g. MitoTracker dye (Em λmax from 516 to 665 nm), CellMask plasma membrane stain available as CellMask™ Orange (Ex λmax 554/Em λmax 567 nm), or Deep red (Ex λmax 649/Em λmax 666 nm). Hoechst or DRAQ5 can be added to living cells to identify the nucleus (6). Following this incubation, the media containing the fluorescent marker is removed and replaced with fresh media.
3. Image cells using laser scanning confocal microscope.

4. Notes

1. The following applies to adherent/semi-adherent cells which will be grown on the glass coverslip. Coverslips are firstly cleaned by soaking in nitric acid for 1 h followed by washing in dH_2O and incubating in methanol overnight. These clean coverslips can then be autoclaved. To improve cell adhesion, the coverslips can be poly-L-lysine coated. Sterile coverslips are placed in a 6-well tissue culture plate and 500 μL of sterile 0.01% poly-L-lysine solution is added to each one and incubated for 5 min at RT. The poly-L-lysine solution can be re-used assuming aseptic technique has been used. The coverslips are then thoroughly rinsed with sterile PBS and then allowed to dry in the hood before addition of cells. Incomplete removal of poly-L-lysine solution results in cell death.
2. The following applies to suspension cells. All incubations on live suspension cells are done in tissue culture flasks or microcentrifuge tubes prior to fixation. Once suspension cells are fixed they can be stored in PBS buffer containing sodium azide (15 mM) for up to 1 month at 4°C without any deleterious effects on the morphology of the cells. Fixed suspension cells are transferred to poly-L-lysine coated coverslips/welled slides. Coverslips/welled slides can be coated by placing in a solution of 0.01% poly-L-lysine for 5 min, removing, and allowing to air-dry. Cells (50–100 μL of a stock solution of 5×10^7 cells/mL) are placed on coated coverslips/welled slides and allowed to attach at room temperature for 30 min. Cells are subsequently processed for microscopy.

3. Fixation methods fall generally into two classes: organic solvents and cross-linking reagents. Organic solvents, such as alcohols and acetone, remove lipids and dehydrate the cells while precipitating the proteins on the cellular architecture. Cross-linking reagents (such as paraformaldehyde) form intermolecular bridges, normally through free amino groups, thus creating a network of linked antigens. Cross-linkers preserve cell structure better than organic solvents, but may reduce the antigenicity of some cell components, and require the addition of a permeabilization step, to allow access of the antibody to the specimen. Here are a couple of alternatives to fixation with paraformaldehyde. The correct choice of method depends on the nature of the antigen being examined and on the properties of the antibody used.

   Methanol Fixation. Fix cells in –20°C methanol for 5–10 min. No permeabilization step needed following methanol fixation.

   Acetone Fixation. Fix cells in –20°C acetone for 5–10 min. No permeabilization step needed following acetone fixation.

   Methanol–Acetone Fixation. Fix in cooled methanol, 10 min at –20°C. Remove excess methanol. Permeabilize with cooled acetone for 1 min at –20°C.

   Methanol–Acetone Mix Fixation. 1:1 Methanol and acetone mixture. Make the mixture fresh and fix cells at –20 C for 5–10 min.

   Paraformaldehyde–Methanol Fixation. Fix in 4% paraformaldehyde for 10–20 min. Rinse briefly with PBS. Permeabilize with cooled methanol for 5–10 min at –20°C.

4. Mounting media containing anti-fade can be made in the lab or bought commercially. Beware, some cyanine dyes react with anti-fade reagents – in particular, media containing aromatic amines, i.e. *p*-phenylenediamine. Aqueous media containing *n*-propyl gallate can be used with cyanine dyes. To prepare aqueous mounting media in the lab, dissolve your anti-fade of choice in PBS containing 50% (w/v) glycerol. Add Hoechst 33342 (Ex λmax 350 nm/Em λmax 461 nm) or DAPI directly to the mounting medium at a final concentration of 1 μg/mL. Both DAPI and Hoechst 33342 stain DNA. The key difference between them is that the ethyl group of Hoechst 33342 renders it more lipophilic and thus more able to cross intact membranes. Hoechst 33342 can be used with fixed or living cells. Numerous aqueous and polymerizing mounting media are available commercially, e.g. ProLong Gold from Invitrogen, VECTASHIELD® from Vector Labs available as an aqueous or a polymerizing media. Ono et al. (7) have written an excellent article on various mounting media.

5. For immunofluorescence studies, lower concentrations of cells ($0.1–0.5 \times 10^6$ cells/mL) can be used, as this allows cells to grow relatively uncrowded, affording better studies of cell morphology. In addition, overcrowding of cells will affect viability and responses to stimuli.
6. The longer the cells are left to incubate on the coverslips just after transfer from culture flasks the more they will spread (making them better for morphological/immunofluorescence studies) and (in the case of cell lines) divide. This should be taken into account when determining the seeding density of the cells.
7. Recommended times and concentrations of wheat germ agglutinin (WGA) may vary in different model systems and require optimization. WGA selectively binds to *N*-acetylglucosamine and *N*-acetylneuraminic acid (sialic acid) residues; therefore, it is a useful membrane marker. However, it tends to stain the plasma membrane of fixed cells better than living cells due to the fact that it binds to receptors in the plasma membrane and is rapidly internalized. Nevertheless, permeabilization of cells following staining does not affect the WGA localization. Another alternative for plasma membrane staining is CellMask™ plasma membrane stain also available from Invitrogen. This can be used on both living and fixed cells and is slow to internalize in live cell imaging. However, the staining with CellMask™ does not survive detergent extraction, and therefore cannot be used in conjunction with probes requiring permeabilization.
8. The concentration and time of incubation of all primary and secondary antibodies should be titrated when being used for the first time as these will vary with cell type and antigen specificity.
9. Alexa Fluors are generally more stable, brighter, and less pH-sensitive than other dyes of comparable excitation and emission. Their fluorescence persists longer than other dyes when exposed to laser excitation.
10. Air bubbles in the mounting media are detrimental to good imaging; therefore, if for example you use prolong gold which is stored at –20°C, ensure that it has equilibrated to room temperature before addition to the sample and apply slowly and with caution.
11. Perfusion is an important step that helps to lower background fluorescent signals that can be caused by the cross reactivity of antibodies in the circulatory system with a secondary antibody of the same species as the tissue, i.e. secondary anti-mouse IgG on a mouse tissue section.
12. Thinner tissue sections work better for immunohistochemistry staining. Thicker sections, e.g. those greater than 15 μm, can

cause problems with permeabilization of various probes. This varies with tissue composition.

13. In order for the cells to last the duration of the imaging experiment, one must ensure that the microscope has adequate heating and CO_2 and an incubator to maintain both throughout. Living specimens should be kept in a medium that is buffered to maintain the correct pH. Many commonly used culture media are buffered with bicarbonate and require an atmosphere with 5–10% CO_2 to maintain the correct pH. Microscope enclosures, stage warmers, temperature-controlled chambers, objective heaters, and CO_2 controllers are available from suppliers of microscopes and microscope accessories, e.g. PeCon GmbH.

References

1. Peng, X. H., Cao, Z. H., Xia, J. T., Carlson, G. W., Lewis, M. M., Wood, W. C., and Yang, L. (2005) Real-time detection of gene expression in cancer cells using molecular beacon imaging: new strategies for cancer research, *Cancer Res 65*, 1909–1917.
2. Bystricky, K., Van Attikum, H., Montiel, M. D., Dion, V., Gehlen, L., and Gasser, S. M. (2009) Regulation of nuclear positioning and dynamics of the silent mating type loci by the yeast Ku70/Ku80 complex, *Mol Cell Biol 29*, 835–848.
3. Palsson-McDermott, E. M., Doyle, S. L., McGettrick, A. F., Hardy, M., Husebye, H., Banahan, K., Gong, M., Golenbock, D., Espevik, T., and O'Neill, L. A. (2009) TAG, a splice variant of the adaptor TRAM, negatively regulates the adaptor MyD88-independent TLR4 pathway, *Nat Immunol 10*, 579–586.
4. Hanrahan, O., Webb, H., O'Byrne, R., Brabazon, E., Treumann, A., Sunter, J. D., Carrington, M., and Voorheis, H. P. (2009) The glycosylphosphatidylinositol-PLC in Trypanosoma brucei forms a linear array on the exterior of the flagellar membrane before and after activation, *PLoS Pathog 5*, e1000468.
5. Helming, L., and Gordon, S. (2007) Macrophage fusion induced by IL-4 alternative activation is a multistage process involving multiple target molecules, *Eur J Immunol 37*, 33–42.
6. Martin, R. M., Leonhardt, H., and Cardoso, M. C. (2005) DNA labeling in living cells, *Cytometry A 67*, 45–52.
7. Ono, M., Murakami, T., Kudo, A., Isshiki, M., Sawada, H., and Segawa, A. (2001) Quantitative comparison of anti-fading mounting media for confocal laser scanning microscopy, *J Histochem Cytochem 49*, 305–312.
8. Harris, J., Schwinn, N., Mahoney, J. A., Lin, H. H., Shaw, M., Howard, C. J., da Silva, R. P., and Gordon, S. (2006) A vitellogenic-like carboxypeptidase expressed by human macrophages is localized in endoplasmic reticulum and membrane ruffles, *Int J Exp Pathol 87*, 29–39.

Chapter 13

Isolation of Exosomes for Subsequent mRNA, MicroRNA, and Protein Profiling

Sweta Rani, Keith O'Brien, Fergal C. Kelleher, Claire Corcoran, Serena Germano, Marek W. Radomski, John Crown, and Lorraine O'Driscoll

Abstract

Exosomes are nano-sized, cell membrane surrounded structures that are released from many cell types. These exosomes are believed to transport a range of molecules, including mRNAs, miRNAs, and proteins; the contents depending on their cell of origin. The physiological and pathological relevance of exosomes has yet to be fully elucidated. Exosomes have been implicated in cell-to-cell communication. For example, in relation to the immune system, such exosomes may enable exchange of antigen or major histocompatibility complex–peptide complexes between antigen-bearing cells and antigen-presenting cells; in cancer, they may contain molecules that not only have relevance as biomarkers, but may also be taken up and cause adverse effects on secondary cells. Furthermore, exosomes have been proposed as autologous delivery systems that could be exploited for personalised delivery of therapeutics. In order to explore the contents and functional relevance of exosomes from medium conditioned by culture cells or from other biological fluids, prior to extensive molecular profiling, they must be isolated and purified. Here, we describe differential centrifugation methods suitable for isolating exosomes from conditioned medium and from other biological fluids, including serum, saliva, tumour ascites, and urine. We also detail Western blotting and transmission electron microscopy methods suitable for basic assessment of their presence, size, and purity, prior to progressing to global mRNA, miRNA, or protein profiling.

Key words: Exosomes, Multivesicular bodies, Extracellular, Cell line, Conditioned medium, Serum, Plasma, Urine, Saliva

1. Introduction

Exosomes are membrane-bound nanoparticles (30–100 nm in diameter) that are released by many cell types. These small, right-side-out structures form intracellularly by inward budding of endosome membranes (1), resulting in vesicles-containing endosomes

Lorraine O'Driscoll (ed.), *Gene Expression Profiling: Methods and Protocols*, Methods in Molecular Biology, vol. 784,
DOI 10.1007/978-1-61779-289-2_13, © Springer Science+Business Media, LLC 2011

called multivesicular bodies (MVBs). When MVBs fuse with the cell membrane, they release their internal vesicles into the extracellular environment. Once released into extracellular space, these microvesicles are termed exosomes (2).

The physiological and pathological role(s) of exosome is of great interest. Exosomes have been implicated in cell-to-cell communication via trans-cellular signalling (3), transfer of membrane receptors, proteins, mRNA, microRNA (miRNA) (4), and organelles (e.g. mitochondria) between cells. Other roles with which exosomes have been associated include the delivery of infectious and toxic agents (e.g. chemotherapeutic drugs) into cells (5). The contents of exosomes depend on their cell of origin (6). Exosomes derived from cells of the immune system enable exchange of antigen or major histocompatibility complex (MHC)-peptide complexes between antigen-bearing cells and antigen-presenting cells. Exosomes have also been described as having both immunostimulatory and anti-tumour affects in vivo (7). Exosomes derived from tumour cells have been associated with accelerating tumour growth (6, 8) and invasiveness (9–11).

Based on the studies of conditioned medium (CM) derived from cell lines and primary cell cultures, as well as from analysis of bodily fluids, including serum/plasma, urine, and saliva, ourselves and others have also reported evidence to suggest that extracellular mRNAs, miRNAs, and proteins may be contained and protected in membrane-bound structures (12, 13). This is supported by the fact that the RNA contained within exosomes remains amplifiable, implicating protection from RNase degradation by the exosome membrane (6, 14, 15). Circulating exosomes have been identified as having potential diagnostic relevance in various cancer types, including ovarian cancer (16), glioblastomas (6), and lung cancer (17). In fact, these structures have been proposed to be involved in horizontal transfer of information between cells, suggesting that mRNAs/miRNAs/proteins could be carried from a cancer cell and be taken up and subsequently cause adverse effects on secondary cells. Intriguingly, emerging data suggest that exosomes may have a role as autologous delivery systems that could be harnessed for personalised delivery of therapeutics into secondary cells.

Following isolation of pure populations of exosomes, these entities are suitable for a range of gene expression profiling approaches in order to determine their contents and so contribute to our understanding of their relevance. Such profiling may include RT-PCR, qPCR, and multiplex PCR (as outlined in Chapters 1 and 2); global mRNA/whole genome microarray analysis (as in Chapter 3); global miRNA analysis (as in Chapter 7); proteomics, including 2D gel electrophoresis, mass spectrometry, and Western blotting (as in Chapters 8 and 9). Thus, detailed protocols for the isolation of exosomes from CM and bodily fluids using centrifugation and subsequent basic characterisation of exosomes (i.e. their presence and purity) are reported in this chapter.

2. Materials

2.1. Cell Culture and Conditioned Medium Collection

1. MCF7 (American Type Tissue Collection) – RPMI-1640, 2-mM L-Glutamine, 5% foetal bovine serum (FBS) from which exosomes have been eliminated (see Notes 1 and 2).
2. Tissue culture grade vented flasks (e.g. 175 cm^2).
3. 30- and 50-mL centrifuge tubes.
4. 0.22-μm filters.
5. 20- and 50-mL syringes.
6. Biosafety cabinet, pipette aids, etc. (as for basic mammalian cell culture).

2.2. Serum Collection

1. Non-heparinised tube(s) for blood procurement (by trained Phlebotomist).
2. Bench-top centrifuge.
3. Cryovial tubes.
4. –80°C freezer.

2.3. Exosome Collection

1. Refrigerated bench-top centrifuge.
2. 30- and 50-mL centrifuge tubes.
3. Ultracentrifuge and fixed-angle or swinging-bucket rotors (for details on Beckman Coulter ultracentrifuge, rotors, and rpm to *g* conversions, see: http://www.beckmancoulter.com/resourcecenter/labresources/centrifuges/rotorcalc.asp).
4. Appropriately sized polyallomer or polycarbonate tubes for the rotor(s) mentioned in Item 3.
5. Phosphate-buffered saline (PBS): sodium chloride 8 g/L, potassium chloride 0.2 g/L, di-sodium hydrogen phosphate 1.15 g/L, and potassium dihydrogen phosphate 0.2 g/L, pH 7.3 at 25°C (Sigma).

   For the additional/option step of exosome purification on sucrose cushion

6. Tris/sucrose/D_2O: 30 g protease-free sucrose, 2.4 g Tris base, and 50 mL D_2O. Adjust pH to 7.4 with 10 N HCl. Bring to 100 mL total volume with D_2O. Pass through a 0.2-μM filter. Store for up to 2 months at 4°C.
7. 5- and 50-mL syringes.

8. 18-G needle.
9. 0.2-μM filter.
10. SW 28 ultracentrifuge rotor and appropriately sized polyallomer tubes.
11. 45 Ti ultracentrifuge rotor and appropriately sized polycarbonate tubes.

2.4. Western Blotting

1. Laemmli sample buffer (2×): 4% SDS, 20% glycerol, 10% 2-mercaptoethanol, 0.004% bromphenol blue, and 0.125 M Tris–HCl, pH 6.8. Stored at –20°C.
2. Resolving buffer (4×): 1.5 M Tris–HCl, pH 8.8, 0.4% SDS. Stored at room temperature.
3. Stacking buffer (4×): 0.5 M Tris–HCl, pH 6.8, 0.4% SDS. Stored at room temperature.
4. Acrylamide/Bis-acrylamide, 30% solution (Sigma). Stored at 4°C.
5. *N,N,N,N′*-Tetramethyl-ethylenediamine (TEMED; Sigma). Stored at 4°C.
6. Ammonium persulfate: 10% solution in double-distilled water, freshly prepared.
7. Running buffer (10×): 250 mM Tris, 1.92 M glycine, 1% (w/v) SDS. Stored at room temperature.
8. Prestained Molecular weight markers: PageRuler™ Prestained Protein Ladder (Fermentas, Burlington, Canada).
9. 1-D Electrophoresis system (Bio-Rad Laboratories Inc., Hercules, CA).
10. Blotting buffer: 25 mM Tris–HCl pH 8.3, 192 mM glycine and 20% (v/v) methanol. Stored at 4°C. Methanol to be added immediately before use.
11. Immun-Blot PVDF membrane and extra thick blot paper (Bio-Rad Laboratories Inc.).
12. Ponceau S solution (Sigma).
13. Tris-buffered saline (TBS, 10×): 100 mM Tris–HCl pH 7.5, 1.5 M NaCl. Stored at room temperature.
14. Blocking buffer: Membranes were blocked in 5% low-fat dry milk (Bio-Rad).
15. Washing buffer (TBS-T): 1× TBS solution supplemented with 1% Tween-20. Stored at room temperature.
16. Antibody dilution buffer: 1× TBS supplemented with 3% (w/v) bovine serum albumin (BSA) and 1% Tween-20. Single aliquots frozen at –20°C.

17. Primary antibody: mouse anti-TSG101 antibody (Abcam): 1:500 dilution in 3% Blotting-Grade Blocker and 0.1% Tween (Sigma).
18. Secondary antibodies: Horseradish peroxidase (HRP) conjugated anti-mouse antibody (Cell Signalling).
19. Enhanced chemiluminescent (ECL) reagents: Immobilon Western Chemiluminescent HRP Substrate (Millipore, Billerica, MA).
20. Stripping buffer: 62.5 mM Tris–HCl, pH 6.8, 2% (w/v) SDS. Stored at room temperature. Warm to 70°C and add 100 mM β-mercaptoethanol before use.
21. Wet electroblotting system (Bio-Rad Laboratories Inc.).
22. Visualisation: Proteins were visualised by chemiluminescence (Millipore).
23. Imaging system: Detection was performed with the Chemidoc exposure system (Bio-Rad Laboratories).

2.5. Transmission Electron Microscopy

1. Transmission electron microscopy (TEM).
2. PBS: Sodium chloride 8 g/L, potassium chloride 0.2 g/L, di-sodium hydrogen phosphate 1.15 g/L, and potassium dihydrogen phosphate 0.2 g/L, pH 7.3 at 25°C (Sigma).
3. 4% paraformaldehyde: Dissolve 4 g paraformaldehyde in 90 mL of 0.1 M sodium phosphate buffer. Heat to 65°C while stirring. Carefully add 1N NaOH, dropwise, until the solution clears. Bring to 100 mL with 0.1 M sodium phosphate buffer, allow to cool and then filter through a 0.22-μm filter. This may be used immediately or frozen (once) and stored at −20°C for up to 6 months.
4. 1% glutaraldehyde: Dilute glutaraldehyde in 0.1 M sodium phosphate buffer, pH 7.4, to a final concentration of 1%. This may be used immediately or frozen (once) and stored at −20°C for up to 6 months.
5. Uranyl acetate, pH 4.0: Dissolve 2 g of uranyl acetate in 50 mL distilled H_2O. This may be stored at 4°C in the dark for up to 6 months. Just prior to use, filter required volume through a 0.22-μm filter.
6. Methyl cellulose, 2%: Dissolve 4 g methyl cellulose (Sigma) in 196 mL of distilled H_2O, which has been heated to 90°C. Stirring will be required to dissolve completely. Continuing to stir while rapidly cooling on ice to 10°C, and then more slowly cool to 4°C overnight while stirring. Leave to rest for 3 day, and subsequently bring to a final volume of 200 mL with water. Centrifuge at 100,000×g for 95 min at 4°C. Collect supernatant 2% methyl cellulose. This may be stored at 4°C for up to 3 months.

7. 0.15 M oxalic acid: dissolve 0.945 g oxalic acid in 50 mL distilled H_2O.
8. Uranyl-oxalate, pH 7.0: Mix uranyl acetate (pH 4.0) with 0.15 M oxalic acid (1: 1). Adjust to pH 7.0 by the dropwise addition of 25% (w/v) NH_4OH. This solution may then be stored, in the dark, at 4°C for up to 1 month.
9. Formvar-coated grids.
10. Parafilm.
11. Whatman no. 1 filter paper.
12. Forceps.
13. Grid storage box.

3. Methods

A number of methods exist for the isolation of exosomes from medium conditioned (CM) by cell lines and from bodily fluids (e.g. serum, urine, etc). As the most commonly used method for this purpose is differential centrifugation, we detail this approach here.

3.1. Collecting CM for Subsequent Exosomes Isolation

1. Grow cells of interest (MCF7 included here, as example) until they reach 70% confluency (see Note 3).
2. Remove the medium and replace with fresh FCS-/exosome-free medium, from which cell MCF7-secreted exosomes are subsequently collected. Incubate for 48 h (see Note 4).
3. Collect the CM supernatant and transfer to 30- or 50-mL polypropylene tubes. Centrifuge at 300 × *g* for 10 min at 4°C to remove any free cells. Carefully collect the CM supernatant with a pipette; do not pour (see Note 5).
4. Transfer the CM into a fresh centrifuge tube and spin at 2,000 × *g* for 20 min at 4°C, to remove large cell particles/cell debris. As mentioned above, carefully collect the CM with a pipette and proceed as outlined in Subheading 3.3.

3.2. Serum Collection for Subsequent Exosomes Isolation

1. Allow blood in non-heparinised tube(s) to clot for 30 min minimum to 1 h maximum after procurement.
2. Centrifuge at 400 × *g* for 15 min.
3. Gently remove the serum and dispense (as approx. 0.5 mL volumes) into labelled cryovial tubes.
4. Cryovial tube label should include: (a) anonymised identifier; (b) date; (c) duration of time from procurement to placing at –80°C. Keep in mind that the time from procurement to placing at –80°C should be <3 h.

3.3. Isolating Exosomes from CM

1. Progressing from Step 4 of Subheading 3.1, transfer the CM into a polyallomer or polycarbonate tube(s) appropriate to the ultracentrifuge rotor to be used (see Note 6).
2. Using a waterproof marker, mark one side of each ultracentrifuge tube and carefully place the tubes into the rotor in such a way that the mark is facing out/up, as a reference to where the pellet will be following centrifugation. For fixed-angle rotors, the pellet will be found near the bottom of the tube and on the side of the tube facing up; for swinging-bucket rotors, the pellet will be at the bottom of the tube.
3. Centrifuge at 10,000 × *g* for 30 min at 4°C.
4. Again, gently removing the CM supernatant, repeat Steps 1 and 2 of Subheading 3.3. Even if a pellet is too small to be visible, assume its position to be as mentioned in Step 2 and leave approximately 0.5 cm of medium behind, to avoid disturbing the pellet and thus contaminating the CM supernatant.
5. *Importantly*, for CM from many cell types, Steps 3 and 4 inclusively can be avoided by passing the CM through a 0.22-μm filter to remove any remaining large cell particles/cell debris (see Note 7).
6. After placing the CM supernatant into fresh, marked ultracentrifuge tubes, spin at 110,000 × *g* for 70 min at 4°C; this time to pellet the exosomes.
7. Carefully remove and discard the supernatant CM, saving the exosome pellet.
8. To wash any protein contamination off the exosome pellet, gently re-suspend the pellet in 1 mL PBS (see Note 8). Pool exosome suspensions from all tubes containing exosomes from the same CM and place into a fresh ultracentrifuge tube. Fill the tube with PBS.
9. Centrifuge at 110,000 × *g* for 70 min at 4°C.
10. Re-suspend the pellet in a small volume (typically 50–100 μL) of PBS.

3.4. Isolating Exosomes from Bodily Fluids

In principle, the method for isolating exosomes from bodily fluids, such as serum, plasma, saliva, broncho-alveolar lavage, tumour asites, etc., by differential centrifugation is similar to that for CM. However, as some of these fluids are more viscous than CM, diluting in PBS as well as increasing the duration and – in some cases – the speed of centrifugation is recommended. As mentioned above, the method outlined here is used successfully for isolating exosomes from serum. However, a similar approach is suitable for procuring exosomes from other bodily fluids.

1. Progressing from Step 4 of Subheading 3.2, dilute the serum with an equal volume of PBS and subsequently bring to a 10 mL total volume with PBS.

2. Pass this suspension through a 0.22-µm filter to remove any large cell particles or cell debris.
3. Transfer the serum supernatant into a 30- or 50-mL centrifuge tube and centrifuge at 2,000 × *g* for 30 min at 4°C.
4. As mentioned above, without disturbing the pellet and so risk avoiding contamination, carefully collect the serum supernatant with a pipette and transfer into a polyallomer or polycarbonate tube appropriate to the ultracentrifuge rotor to be used (see Note 6). As outlined in Step 2 of Subheading 3.3, mark the ultracentrifuge tubes and subsequently centrifuge at 12,000 × *g* for 45 min at 4°C. *Importantly*, as an alternative to this spin at 12,000 × *g*, at this stage the serum supernatant can be passed through a 0.22-µm filter.
5. Place the resulting serum supernatant into a fresh tube, and centrifuge at 110,000 × *g* for 2 h at 4°C to pellet the exosomes.
6. Carefully remove the supernatant, this time saving the exosome pellet.
7. To wash the exosome pellet, gently re-suspend in 1 mL PBS (see Note 8). Where relevant, pool exosome suspensions from all tubes containing exosomes from the same serum specimen and place into a fresh centrifuge tube. Fill the tube with PBS.
8. Centrifuge at 110,000 × *g* for 70 min at 4°C.
9. Re-suspend the washed exosomes pellet in a small volume (typically 50–100 µL) of PBS.

3.5. Further Purification of Isolating Exosomes (Optional)

The procedures outlined above are generally adequate for exosome isolation. However, prior to progressing to extensive characterisation, it is recommended that the purity of the initial exosome population isolated from a given source be assessed, e.g. by using electron microscopy. If the population is found not to be of an acceptable purity (e.g. if it includes large protein aggregates), an additional step involving further purification of the exosomes on a sucrose gradient is recommended. To do so:

1. Gently re-suspend the exosome pellet (for CM, from Step 9 of Subheading 3.3; for serum, from Step 8 of Subheading 3.4 in 25 mL PBS).
2. Place 4 mL of Tris/sucrose/D_2O solution into a polyallomer tube (appropriate for SW 28 ultracentrifuge rotor).
3. Carefully layer the diluted exosomes onto the cushion of sucrose without mixing.
4. Centrifuge at 110,000 × *g* for 70 min at 4°C.
5. Using the 18-G needle on the 5-mL syringe, pierce the lower part of the tube and gently draw off ~3.5 mL of the Tris/sucrose/D_2O solution that now contains the exosomes.

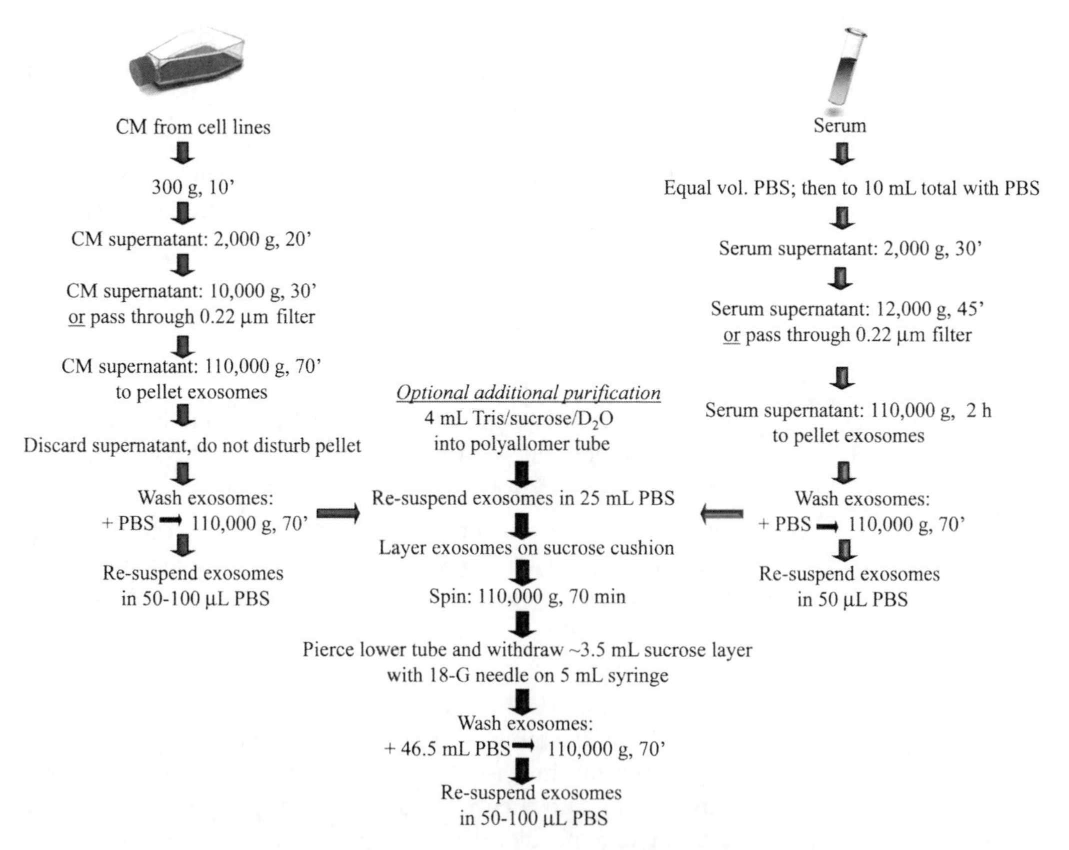

Fig. 1. Schematic representation of exosomes isolation by differential centrifugation. Exosomes can be isolated from medium conditioned by cell lines or primary cultures and from serum, using the procedures illustrated here. If sample analysis of the exosome isolates (e.g. by transmission microscopy) indicates an impure preparation, an additional step of purifying the exosomes on a sucrose cushion is recommended. All centrifugation steps should be performed at 4°C.

6. To wash the exosomes of sucrose, transfer this solution to a fresh ultracentrifuge tube; dilute to 50 mL with PBS; and centrifuge at 100,000 × *g* for 70 min at 4°C in a 45 Ti rotor.
7. Re-suspend the pellet in a small volume (typically 50–100 µL) of PBS (see Fig. 1 *for summary of exosome isolation procedures*).

3.6. Basic Characterising of Exosomes

Prior to progressing with extensive mRNA/miRNA/protein profiling, it is recommended that samples of exosome isolates be assessed for presence, size, and purity. While there are a range of possible options, typically Western blotting is used to show the presence of common exosomal proteins (e.g. Alix, 96 kDa; Tsg101, 44 kDa; CD9, 25 kDa) and TEM can be used to show the presence, size – and to some extent, purity – of exosomes isolates.

3.6.1. Western Blotting

For protein isolation and quantification prior to Western blotting analysis, techniques as outlined in Chapter 8 may be used.

1. The protocol here reported refer to the use of mini gels 1-D Electrophoresis system (Bio-Rad Laboratories), but can be adapted to other formats.
2. Carefully clean glass plates with 95% (v/v) ethanol and assembly the front and back glasses in the opposite clamps. It is advisable to pour dH_2O inside the plates to check that there is no leakage from the bottom of the set-up. Then pour off the dH_2O.
3. Prepare a 7.5% gel solution by mixing 2 mL of 4× resolving buffer and 2 mL of 30% acrylamide/bis-acrylamide solution with 4 mL of water. Subsequently, add 100 μL of ammonium persulfate solution and 40 μL of TEMED; mix and immediately pour the gel, leaving enough space for the stacking gel. Overlay with ethanol and allow to polymerise.
4. Pour off the ethanol and rinse with water. Then pour the stacking gel solution prepared by mixing 1 mL of 4× stacking buffer, 0.5 mL of 30% acrylamide/bis-acrylamide solution, and 2.5 mL of water. Immediately insert the combs and allow to polymerise.
5. Carefully remove the glass plates containing the gel from the holder and assemble the gasket with the electrodes. Remove the comb, fill the gasket and the outer chamber with running buffer (1×), and then wash the wells with a syringe fitted with a thin gauge needle.
6. Load the samples and the molecular markers in the wells.
7. Assemble the unit and connect to the power supply. Apply a constant voltage up to 130 V to carry samples through the stacking gel, and then increased to 150 V to run through the running gel. Turn off the power supply immediately after the bromophenol blue dye has run off the gel and disconnect the power supply.
8. Cut a sheet of PVDF paper to a size that is slightly larger than the gel size and place in a tray containing methanol to activate the membrane. After 1 min, transfer the PVDF in another tray filled with blotting buffer. Wet two sheets of extra thick paper and two sponges in blotting buffer.
9. Disassemble the gel unit, cut and remove the stacking gel with a blade, then transfer the resolving gel into a tray containing blotting buffer. If desired, a corner of the gel can be cut to allow the tracking of the gel orientation.
10. Assemble the transfer cassette by lying a sheet of paper onto a sponge and the PVDF membrane on the top; the gel is then carefully laid on top of the membrane; another sheet of thick paper and a sponge are then positioned on top of the gel. Make sure that no air-bubbles exist between the gel and the membrane. The transfer cassette is then locked.

11. Insert the cassette in the transfer tank, carefully checking the orientation, the membrane must be oriented towards the anode, while the gel towards the cathode. Put an iced-freezer pack in the tank, add enough blotting buffer to cover the cassette and activate a magnetic stir-bar in the tank to avoid heating of the buffer.
12. Close the lid, connect the unit to the power supply and begin transfer, with a constant current of 200 mA for 1.5 h.
13. Following this, disconnect the power supply and then disassemble the transfer cassette. Remove the sponge, the paper, and the gel. If using prestained markers, check that their corresponding bands are clearly visible on the membrane.
14. Place the membrane in a small dish and add 10 mL of Ponceau S staining solution and incubate the membrane for 2 min. The Ponceau S solution can be reused several times.
15. Wash the membrane briefly with TBS to remove excess staining. Continue washing the membrane until the staining is gone. If stains persist, wash the membrane with TBS containing 0.02% NaAzide for 1–3 min and then rinse once with TBS.
16. The membrane is then incubated in 15 ml blocking buffer for 1 h at room temperature on a rocker with gentle shaking.
17. After blocking, the membrane is rinsed twice with TBS and then incubated for 3 h with primary antibody solution at room temperature with gentle shaking. Alternatively, the incubation can be performed overnight at 4°C to enhance the signal.
18. The primary antibody is then removed and the membrane is washed three times for 10 min each with 15 ml of TBS-T with vigorous shaking.
19. Freshly prepared secondary antibody solution is then added to the membrane for 1 h at room temperature with gentle shaking.
20. The secondary antibody solution is discarded and three washes for 10 min each with TBS-T are performed with vigorous shaking.
21. The ECL reagents are mixed together at a ratio of 1:1 immediately before use and evenly added to the blot for 3 min.
22. The excess ECL is removed and the membrane is put in a tray. Proceed with image acquisition as detailed in the next section.
23. Once a satisfactory signal has been obtained, wash the membrane and then proceed with the stripping procedure to clear the membrane before re-probing for a housekeeping gene.
24. For the stripping procedure, warm 30 ml stripping buffer at 70°C and then add. Incubate the membrane in this solution

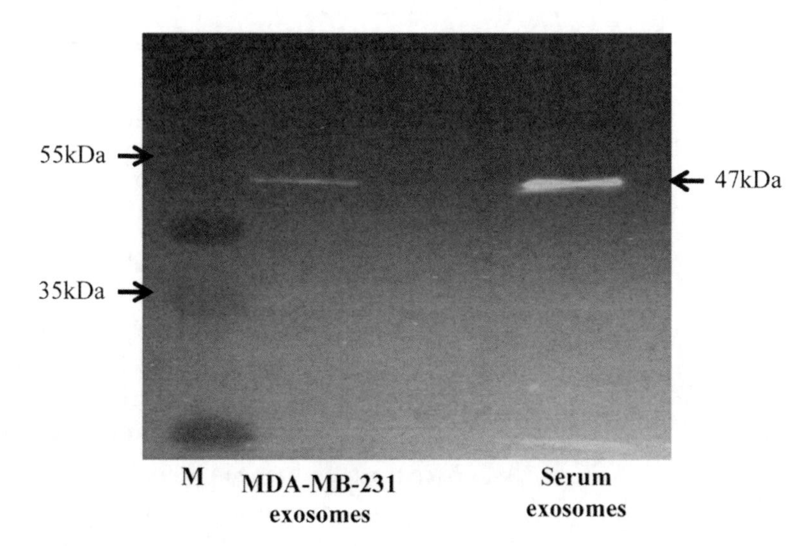

Fig. 2. Western blot analysis of samples of exosome isolates. Probing for tumour susceptibility gene 101 (TSG101), a protein typically assessed as an exosome marker, showed successful isolation of exosomes from cell line conditioned medium (CM: example shown is MDA-MB-231) and from human serum.

for 30 min and then perform extensive washes with TBS. Repeat the blocking step again before re-probing with the primary antibody solution.

25. Detection and imaging can be performed using a Chemidoc exposure system, as outlined in Chapter 8, Subheading 3.4 (see Fig. 2 e.g. *TSG101 Western blot*).

3.6.2. Transmission Electron Microscopy

The procedure outlined here is an adaptation of a technique previously reported, which can be successfully used for analysis of exosomes (18).

1. Using 4% paraformaldehyde, re-suspend exosomes that have been pelleted at 110,000 × *g* at 4°C. Final volume should be ≤100 μL (see Note 9).
2. Place a drop (~10 μL) of this suspension onto a small sheet of parafilm and invert a formvar-coated EM grid onto this, allowing the grid to float on the suspension for 20 min (see Note 10).
3. Gently wash the exosomes-containing grid in the same way, i.e. transfer onto a drop (50 μL) of PBS and leave for 1 min. Repeat this step by transferring onto fresh drops of PBS; 2× more times.
4. Fix the exosomes-containing grid by transferring onto 50 μL of 1% glutaraldehyde for 5 min.
5. Transfer the exosomes-containing grid onto 100 μL distilled H_2O for 2 min. Repeat this step by transferring onto fresh drops of PBS; 4× more times.

6. Place grid on 50 μL of uranyl-oxalate pH 7.0, for 5 min.
7. Transfer onto 50 μL of methyl cellulose-uranyl acetate (9:1 mix of 2% methyl cellulose:4% uranyl acetate); for 10 min on ice.
8. Again and as outlined in Note 10, remove excess methyl cellulose-uranyl acetate solution from the grid by gently touching the side (circumference) of the grid against Whatman no. 1 filter paper.
9. Allow the exosomes-containing grid to air-dry for approximately 10 min (see Note 11).
10. When dry, observe under EM or store in grid box until ready for analysis.

4. Notes

1. MCF7 is included here as example cell line. Of course, this approach can be applied to other cell lines of choice, cultured in their optimal medium.
2. Exosomes are present in FCS. For this reason, the following recommendations should be considered to avoid "contaminating" the exosomes of interest with exosomes arising from the FCS. If the cells of interest can be maintained under serum-free conditions, this is the most straightforward approach. If the cells need some protein to survive, 1% (w/v) BSA may be added instead of FCS. If cells cannot survive without FCS, it is recommended that serum used be depleted of exosomes (as in Steps 3–6 of Subheading 3.4, but keep the serum in this instance, rather than the FCS exosomes), prior to its addition to culture medium. However, as this can be a quite laborious additional step, some researchers choose to use a limited amount of whole FCS (e.g. 1%), but must accept that a small amount of exosomes subsequently isolated may be arising from the FCS.
3. Use as many flasks/"units" of cells as necessary to produce a minimum of 70–100 mL CM. As the yield from the purification procedure increases with the starting volume, it is advisable to purify exosomes from large volumes of CM.
4. Cells must be in a healthy, non-apoptotic stage. For some cell types that grow rapidly, cells may become over-confluent and start to die by 48 h. In this case, 24-h incubation may be more appropriate. For other cell lines that expel exosomes in a limited way, 72-h incubation may be ideal. The timing should be optimised for each cell line of interest.
5. Once CM is collected, it is strongly recommended to proceed immediately with exosomes isolation. However, this is not always

possible (and particular so in the case where human serum specimens are being procedure over time; Subheading 3.2). CM containing exosomes may be stored for a limited number of days at 4°C. Exosomes have been successfully isolated from CM, serum, and other bodily fluids may be stored at –80°C for at least several months. While direct comparisons of immediately analysed versus short-term stored versus long-term stored exosomes have not yet been reported, this storage is likely to lead to some loss of exosomes and the resulting exosomes may be of more limited use (depending on the intended follow-on experiments).

6. To avoid contaminating the exosomes that are to be subsequently isolated, very carefully remove the CM with a pipette, leaving behind approximately a centimetre of medium above the cell pellet. Do not pour off the supernatant as the pellet may become completely or partially dislodged and contamination would then be unavoidable.
7. It is recommended that the initial centrifugation steps and filtration (through 0.22-μm filter) be assessed with CM from any given cell line to determine the yield of exosomes prior to selecting the optimal method for further studies.
8. Exosome pellets are often so small (due to the nano-size of exosomes) that they are not visible to the naked eye. However, from the marking that you have placed on the ultracentrifuge tube, you will know where the small pellet will have formed and so progress accordingly with care.
9. Exosomes fixed in paraformaldehyde may be stored for up to 1 week prior to proceeding to EM analysis.
10. For this and subsequent washing and fixing steps, ensure that the formvar-coated grid containing the exosomes is not allowed to dry out, while keeping the back of the grid dry. Remove excesses of each solution from the grid, prior to progressing to the next step, by gently touching the side (circumference) of the grid against Whatman no. 1 filter paper.

Acknowledgements

Preparation of this chapter was supported by Trinity College Dublin's Start-Up Funds for New Academics 2008/2009; Trinity College Dublin's Postgraduate Studentship; Science Foundation Ireland's funding of Molecular Therapeutics for Cancer, Ireland [08/SRC/B1410] and the Marie Keating Foundation PhD Scholarships at Trinity College Dublin.

References

1. Théry, C., Zitvogel, L., Amigorena, S. (2002) Exosomes: composition, biogenesis and function. *Nat Rev Immunol.* **2**, 569–79.
2. Keller, S., Sanderson, M.P., Stoeck, A., Altevogt, P. (2006) Exosomes: from biogenesis and secretion to biological function. *Immunol Lett.* **107**, 102–8.
3. Calzolari, A., Raggi, C., Deaglio, S., Sposi, N.M., Stafsnes, M., Fecchi, K., Parolini, I., Malavasi, F., Peschle, C., Sargiacomo, M., Testa, U. (2006) TfR2 localizes in lipid raft domains and is released in exosomes to activate signal transduction along the MAPK pathway. *J Cell Sci.* **119**, 4486–98.
4. Valadi, H., Ekström, K., Bossios, A., Sjöstrand, M., Lee, J.J., Lötvall, J.O. (2007) Exosome-mediated transfer of mRNAs and microRNAs is a novel mechanism of genetic exchange between cells. *Nat Cell Biol.* **9**, 654–59.
5. Safaei, R., Larson, B.J., Cheng, T.C., Gibson, M.A., Otani, S., Naerdemann, W., Howell, S.B. (2005). Abnormal lysosomal trafficking and enhanced exosomal export of cisplatin in drug-resistant human ovarian carcinoma cells. *Mol Cancer Ther.* **4**, 1595–1604.
6. Skog, J., Würdinger, T., van Rijn, S., Meijer, D.H., Gainche, L., Sena-Esteves, M., Curry, W.T. Jr., Carter, B.S., Krichevsky, A.M., Breakefield, X.O. (2008) Glioblastoma microvesicles transport RNA and proteins that promote tumour growth and provide diagnostic biomarkers. *Nat Cell Biol.* **10**, 1470–76.
7. Chaput N, Taïeb J, André F, Zitvogel L. (2005) The potential of exosomes in immunotherapy. *Expert Opin Biol Ther.* **5**, 737–747.
8. Liu, C., Yu, S., Zinn, K., Wang, J., Zhang, L., Jia, Y., Kappes, J.C., Barnes, S., Kimberly, R.P., Grizzle, W.E., Zhang, H.G. (2006) Murine mammary carcinoma exosomes promote tumor growth by suppression of NK cell function. *J Immunol.* **176**, 1375–85.
9. Ginestra, A., La Placa, M.D., Saladino, F., Cassarà, D., Nagase, H., Vittorelli, M.L. (1998) The amount and proteolytic content of vesicles shed by human cancer cell lines correlates with their in vitro invasiveness. *Anticancer Res.* **18**, 3433–37.
10. Clayton, A., Mitchell, J.P., Court, J., Mason, M.D., Tabi, Z. (2007) Human tumor-derived exosomes selectively impair lymphocyte responses to interleukin-2. *Cancer Res.* **67**, 7458–66.
11. Friel, A.M., Corcoran, C., Crown, J., O'Driscoll, L. (2010) Relevance of circulating tumor cells, extracellular nucleic acids, and exosomes in breast cancer. *Breast Cancer Res Treat.* **123**, 613–25.
12. Hasselmann, D.O., Rappl, G., Tilgen, W., Reinhold, U. (2001) Extracellular tyrosinase mRNA within apoptotic bodies is protected from degradation in human serum. *Clin Chem.* **47**, 1488–89.
13. Ng, E.K., Tsui, N.B., Lam, N.Y., Chiu, R.W., Yu, S.C., Wong, S.C., Lo, E.S., Rainer, T.H., Johnson, P.J., Lo, Y.M. (2002) Presence of filterable and nonfilterable mRNA in the plasma of cancer patients and healthy individuals. *Clin Chem.* **48**, 1212–17.
14. O'Driscoll, L., Kenny, E., Perez de Villarreal, M., Clynes, M. (2005) Detection of Specific mRNAs in Culture Medium Conditioned by Human Tumour Cells: Potential for New Class of Cancer Biomarkers in Serum. *Cancer Genomics & Proteomics* **2**, 43–52.
15. O'Driscoll, L., Kenny, E., Mehta, J.P., Doolan, P., Joyce, H., Gammell, P., Hill, A., O'Daly, B., O'Gorman, D., Clynes, M. (2008) Feasibility and relevance of global expression profiling of gene transcripts in serum from breast cancer patients using whole genome microarrays and quantitative RT-PCR. *Cancer Genomics Proteomics.* **5**, 94–104.
16. Taylor, D.D., Gercel-Taylor, C. (2008) MicroRNA signatures of tumor-derived exosomes as diagnostic biomarkers of ovarian cancer. *Gynecol Oncol.* **110**, 13–21.
17. Rabinowits, G., Gerçel-Taylor, C., Day, J.M., Taylor, D.D., Kloecker, G.H. (2009) Exosomal microRNA: a diagnostic marker for lung cancer. *Clin Lung Cancer.* **10**, 42–46.
18. Théry, C., Clayton A, Amigorena S, Raposo G (2006) Isolation and characterization of exosomes from cell culture supernatants and biological fluids. *In*: Curr Prot Cell Biol 3.22.1–3.22.29 (Wiley & Sons, Inc.).

Chapter 14

Atomic Force Microscopy and High-Content Analysis: Two Innovative Technologies for Dissecting the Relationship Between Epithelial–Mesenchymal Transition-Related Morphological and Structural Alterations and Cell Mechanical Properties

Stephen T. Buckley, Anthony M. Davies, and Carsten Ehrhardt

Abstract

Epithelial–mesenchymal transition (EMT) is a complex series of cellular reprogramming events culminating in striking alterations in morphology towards an invasive mesenchymal phenotype. Increasingly, evidence suggests that EMT exerts a pivotal role in pathophysiological situations including fibrosis and cancer. Core to these dynamical changes in cellular polarity and plasticity is discrete modifications in cytoskeletal structure. In particular, newly established actin-stress fibres supplant a preceding system of highly organised cortical actin. Although cumulative studies have contributed to elucidation of the detailed signalling pathways that underpin this elaborate molecular process, there remains a deficiency regarding its precise contribution to cellular biomechanics. The advent of atomic force microscopy (AFM) and high-content analysis (HCA) provides two innovative technologies for dissecting the relationship between EMT-related morphological and structural alterations and cell mechanical properties. AFM permits acquisition of high resolution topographical images and detailed analysis of cellular viscoelasticity while HCA facilitates a comprehensive and perspicacious assessment of morphological changes. In combination, they offer the possibility of novel insights into the dynamic traits of transitioning cells. Herein, a detailed protocol describing AFM and HCA techniques for evaluation of transforming growth factor-β1-induced EMT of alveolar epithelial cells is provided.

Key words: Cytoskeleton, Epithelial–mesenchymal transition, Atomic force microscopy, High-content analysis microscopy

1. Introduction

Epithelial–mesenchymal transition (EMT) is a manifestation of epithelial plasticity denoted by enduring morphological and molecular alterations in epithelial cells resulting from transdifferentiation to a mesenchymal cell type. This phenotypic conversion

Lorraine O'Driscoll (ed.), *Gene Expression Profiling: Methods and Protocols*, Methods in Molecular Biology, vol. 784,
DOI 10.1007/978-1-61779-289-2_14, © Springer Science+Business Media, LLC 2011

from an epithelial to mesenchymal cell type is a vital element of normal organ development (1). However, it is now widely held, the EMT serves as an important process in the genesis of myofibroblasts in numerous pathological states, including cancer and fibrosis (2–4).

Underlying this dysregulated, interconversion between epithelial and mesenchymal phenotypes is a conspicuous re-organisation of their cytoskeletal systems. Specifically, epithelial cells relinquish their polarised state coupled with disintegration of intercellular junctions, as the actin cytoskeleton undergoes dramatic remodelling from cortical actin to actin-stress fibres (5). Cumulative studies have provided detailed understanding of the contextual cues and molecular mediators that regulate those cellular changes observed during EMT (6, 7). However, there remains a dearth of information regarding the functional contribution of EMT and its associated morphological and structural alterations on cell mechanical properties (8). In this regard, measuring the viscoelastic properties of transitioning living cells offers the possibility of providing novel insights into the influence of EMT-associated cytoskeletal restructuring on cellular biomechanics, and reveal to what extent the elastic properties are caused by cellular components, particularly parts of the cytoskeleton.

Use of two evolving microscopic techniques, namely, atomic force microscopy (AFM) and high-content analysis (HCA) microscopy, permits dual analysis of the elastic properties and morphological traits of cell populations in vitro. Of recent times, AFM has been established as a powerful method in molecular biology, enabling visualisation of cell surface features at the nanoscale, and assessment of cellular mechano-elastic properties – both traditionally difficult properties to characterise (9). Uniquely, such analyses can be performed under native physiological conditions, thus ensuring maintenance of biological systems and their functionality. As a result, it creates the potential for yielding novel insights into countless dynamic molecular events. HCA is a visual biological technology, which facilitates acquisition of spatially and/or temporally resolved data relating to discrete molecular events, and their subsequent quantitative analysis (10). In the context of cell morphology, it allows parallel acquisition of information relating to multiple properties (e.g. cell roundness, cell area, and cell spreading). As such, it represents an automated, unbiased, and rapid analytical method, which enables quantitative evaluation of subtle morphological changes in a statistically robust manner.

In this chapter, we provide a detailed protocol to comprehensively quantify morphological changes induced in alveolar epithelial cells by transforming growth factor (TGF)-β1 stimulation using HCA. Complimentary to this, AFM methods employed to examine EMT-associated cellular topographical features and mechanical stiffness are described.

2. Materials

2.1. Cell Culture

1. Dulbecco's modified Eagle's medium/Nutrient F-12 Ham (Sigma-Aldrich, Dublin, Ireland) containing 15 mM HEPES buffer, 5 mM L-glutamine, 1.2 g/l sodium bicarbonate and supplemented with 5% foetal bovine serum (FBS) (Sigma-Aldrich), 100 U/ml penicillin, and 100 μg/ml streptomycin.
2. Solution of trypsin (0.5 g/l) and ethylenediaminetetraacetic acid (EDTA) (0.2 g/l) from Sigma-Aldrich.
3. Coating solution is prepared by adding 25 μl collagen type I solution (4 mg/ml in 20 mM acetic acid) into 5 ml sterile, tissue culture grade water.
4. TGF-β1 (Peprotech, London, UK) dissolved in 10 mM citric acid giving a stock solution of 50 μg/ml. Aliquots of a working solution (1 μg/ml) are prepared by diluting in PBS containing bovine serum albumin (BSA) (2 mg/ml) and stored at –80°C until use.
5. Lab-Tek™ II Chamber Slides (8-well) (Nunc, Roskilde, Denmark), MicroWell™ plates (96-well) (Nunc), cell culture-treated 6-well plates (Greiner Bio-One, Dublin, Ireland), and circular glass coverslips (24 mm in diameter).

2.2. Cell Fixation and Immunofluorescence Staining for High-Content Analysis Microscopy

1. Dulbecco's phosphate buffer solution (PBS) (Sigma) containing 0.2 g/l KCl, 0.2 g/l KH_2PO_4, 8.0 g/l NaCl, and 1.15 g/l Na_2HPO_4.
2. Paraformaldehyde solution 8% (w/v) in PBS, freshly prepared.
3. Triton X-100 (0.1% v/v) in PBS. Single aliquots frozen at -20°C.
4. BSA 1% (w/v) in PBS, freshly prepared.
5. TRITC-phalloidin (0.1 mg/ml) dissolved in DMSO. Stored at -20°C.
6. Hoechst 33342 (Invitrogen, Germany) dissolved in DMSO giving a stock solution of 10 mg/ml. Single aliquots frozen at -20°C.

2.3. Atomic Force Microscopy

1. For AFM studies on live cells, SiN cantilevers with a nominal spring constant of 0.06 N/m (DNP; Veeco Instruments, Santa Barbara, CA, USA) are used.
2. For AFM imaging on fixed cells, PPP-FM tips (NanoSensors, Neuchatel, Switzerland) with a nominal spring constant of 2.8 N/m are used.
3. To maintain live cells at 37°C, a temperature-controlled liquid cell (BioCell™) is used.

3. Methods

3.1. Cell Culture

The A549 cell line (American Type Culture Collection, ATCC CL-185) is derived from an adenocarcinoma of the lung and exhibits some morphologic and biochemical features of the human pulmonary alveolar type II cell in situ (11). A549 cells contain multilamellar cytoplasmic inclusion bodies, like those typically found in human lung ATII cells, but which disappear with increasing culture time. Notwithstanding their deficiencies, A549 cells remain a widely used model in studies pertaining to the alveolar epithelium.

1. A549 cells are grown in T-75 cell culture flasks and sub-cultured when approaching confluence using trypsin/EDTA to provide new maintenance cultures.
2. In the case of six-well plates incorporating glass coverslips and chamber slides, a seeding density of 10,000 cells/cm^2 is used. For 96-well plates, 2,500 cells are added per well (see Note 1). Cells are cultured at 37°C in 5% CO_2 atmosphere.
3. The amount of collagen I solution to coat six-well plates (35 mm in diameter) containing glass coverslips (24 mm in diameter) is 2 ml, chamber slides 0.4 ml, and 96-well plates (1 mm in diameter) is 100 μl. In each case, allow the collagen to bind for several hours at room temperature after which the excess fluid is removed and each well washed twice with PBS.
4. Following 1 day in culture, the cell medium is removed and replaced with low-serum containing (1%) medium to which TGF-β1 (5 ng/ml) is added to induce EMT (see Note 2).

3.2. Cell Fixation and Immuno-fluorescence Staining

The cytoskeleton is an extremely dynamic structure which is sensitive to alterations in its chemical and mechanical environments. Thus, it is essential that optimal fixation and staining procedures are employed. Fixation facilitates stabilisation of sub-cellular morphological components while preventing degradation of antigens during staining. In this regard, paraformaldehyde, which ensures preservation of a cell's native structure, is preferred. Staining of the actin cytoskeleton is performed using phalloidin, a phallotoxin isolated from the *Amanita phalloides* mushroom. It is capable of staining F-actin at nanomolar concentrations, and represents an efficient fluorescent probe for labelling, identifying, and quantifying F-actin in cellular structures (12).

1. Fix A549 cells (grown in 96-well plates as described in Subheading 3.1) by slowly adding an equal volume of pre-warmed (37°C) 8% paraformaldehyde to culture medium for 15 min at 37°C (see Note 3).

2. Wash twice using PBS.
3. Permeabilise the cells using Triton X-100 (0.1%) in PBS for 5 min.
4. Wash the cells three times using PBS.
5. Incubate the cells with TRITC-phalloidin (final concentration – 500 ng/ml) to visualise filamentous actin and Hoechst 33342 (final concentration – 1 ng/ml) to counterstain cell nuclei at 37°C for 30 min.
6. Wash the cells three times using PBS containing BSA (1%). Then, resuspend in PBS and store at 4°C in the dark until further analysis (see Note 4).

3.3. Image Acquisition Using High-Content Analysis Microscopy

By integrating digital microscopy with software-based image analysis, HCA facilitates the extraction of highly detailed multiparametric cellular information, permitting the analysis of cells at the population and sub-population levels. HCA represents a step change in cellular analysis as it permits the simultaneous monitoring of multiple biological targets, providing detailed information regarding the temporal and spatial distribution of organelles and macromolecules, as well as morphometric and kinetic outputs, thus enabling quantitative denotation of physiological and pathological states.

1. The protocol here reported refers to the use of an IN Cell™ Analyzer 1000 Cellular Imaging and Analysis System (GE Healthcare, Piscataway, NJ). However, it should be possible to adapt to other high-content imaging systems.
2. Allow the 96-well plate(s) to equilibrate to room temperature so as to avoid formation of condensation on the underside of plates, which can result in poor image quality. Any condensation should be removed using a lens cleaning tissue.
3. Switch on the instrument using the on/off switch on the service cabinet.
4. Manually, select the appropriate dichroic filter (e.g. DFTR) to secure the needed excitation–emission spectra for the fluorochromes used.
5. Select the 20× objective lens by turning the turret until it clicks in place.
6. Switch on the lamp power supply and allow to warm-up for 15 min.
7. Dock the 96-well plate into the microscope, positioning well A1 in the top left corner.
8. Load the IN Cell™ Analyzer 1000 Instrument software.
9. To establish a new acquisition protocol, run the "Acquisition Protocol Manager."
10. Under "Objective" specify 20×.

11. Under "Well Plate", specify Nunc MicroWell™ plate (96-well).
12. Using "Well Definition", designate wells as either active or blank according to the experimental design, and define the origin well (i.e. the beginning point for image acquisition).
13. Under "Microscopy", select fluorescence imaging mode and the number of wavelengths (i.e. 2). Brightfield imaging mode should also be selected in order to provide a light microscopical image of the cultured cells.
14. Next, the relevant excitation and emission filter should be selected for each fluorochrome under examination. In the case of TRITC-phalloidin, an excitation of 535 nm and emission of 620 nm should be chosen; for Hoechst 33342, an excitation of 360 nm and emission of 460 nm is required. In addition, a suitable exposure (in milliseconds) should be inputted for each channel, as determined using the auto-focus function.
15. Perform auto offset in order to ensure the optimum focal plane for image acquisition. Searching in 5-μm steps, the microscope will identify the most appropriate distance above the plastic-PBS interface of the sample plate at which the images should be acquired.
16. Following this, specifications regarding the image acquisition are selected. Approximately 15 image fields, which are uniformly distributed throughout the well, should be chosen.
17. Using the software auto-focus feature, attain the *Z*-axis position with the sharpest focus.
18. Next, enter the "Start Acquisition Session Wizard" and select the newly established protocol.
19. Ensure that the correct objective is selected and that the schematic diagram of the microplate layout reflects that of the experimental design.
20. Finally, commence the acquisition process by clicking "Start."

3.4. Morphological Analysis of Acquired Images

This section describes the use of appropriate analytical techniques following image acquisition.

1. The IN Cell™ Analyzer 1000 Workstation software is used to perform analysis of the acquired images.
2. To devise a new analysis protocol, run the "Analysis Protocol Manager."
3. From the Assay name menu, select "Morphology I", the morphology analysis module, and set the microscopy type to fluorescence.
4. In the Images window, assign a wave number (corresponding to the channels representing Hoechst 33342 and TRITC-phalloidin) to each object type (i.e. nuclei and cells).

5. Next, the parameters used to determine image segmentation must be defined (i.e. partitioning the image into a series of defined objects or regions). For nuclei, the "Top Hat" method is used. To obtain a minimum area measurement from a representative image, use the "Arrow" Image Tool. The sensitivity of nuclear detection should be set to 75. Due to the varying size of cell bodies, the "Multiscale Top Hat" method is used in the case of cells (see Note 5). As for nuclei, both the minimum area measurement (using the "Arrow" Image Tool) and sensitivity (60) should be inputted (see Fig. 1).
6. In order to ensure precise analysis of single cells, a series of filters should be employed. In the Filters window, select "Decision Tree" and include filters to exclude irregularly defined nuclei, unusually bright nuclear shapes and an intensity filter for bright shapes in the red wavelength ("Cells").
7. Within the Measures window, the desired measures to acquire should be selected. These should include: 1/(form factor), Cell area, Cell gyration radius, Cell/nuclear area, Nuclear displacement, IxA (N+C), Intensity CV, and Intensity spreading.
8. In order to evaluate an image stack using this analysis protocol, select "Assay Analysis" from the Mode menu and open the desired image stack. Analysis can then be performed on the entire plate or selected wells of interest.
9. The acquired results are then exported to MS Excel and the appropriate data analysis can be performed.

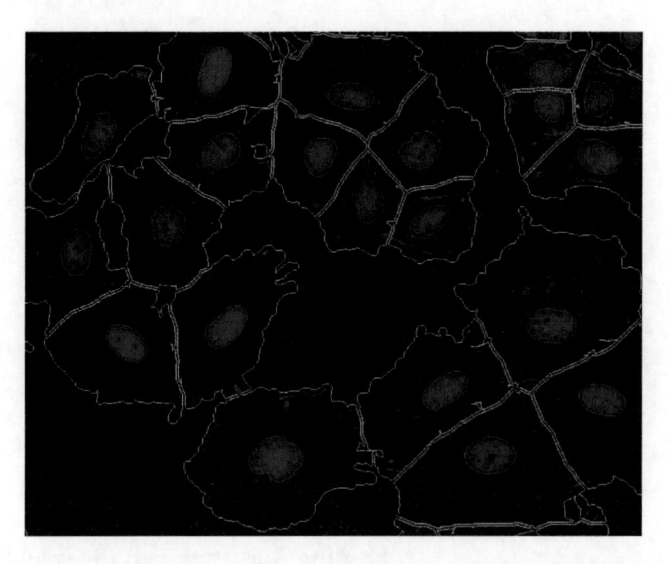

Fig. 1. Morphological analysis of A549 cells using high-content analysis microscopy. Image with analysis bitmap overlay showing results obtained using the morphology analysis protocol described in Subheading 3.4. The image is partitioned into two defined regions, "nuclei" (*blue outline*) and "cells" (*green outline*), using the Top Hat and Multiscale Top Hat methods, respectively.

3.5. Force Measurements Using Atomic Force Microscopy

AFM is well-equipped for measuring nanomechanical properties (e.g. cell stiffness) of biological samples under near physiological conditions. Using contact mode, the tip indents the sample followed by retraction away from the sample. This results in the production of a force–distance curve from which the sample's mechanical features can be determined.

1. The procedure here described employs a NanoWizard II (JPK Instruments, Berlin, Germany) combined with a Nikon Eclipse Ti-E inverted microscope (Nikon Instruments, Surrey, UK), but can be adapted for other makes and models.
2. To ensure optimal performance, the AFM should be enclosed in acoustic and vibration isolation hood.
3. Using a tweezers place, the DNP cantilever tip (Veeco Instruments, Santa Barbara, CA, USA) onto the glass holder and lock in place using the spring provided. Next, fix the glass head in place in the AFM head.
4. Insert the glass coverslip containing A549 cells into the BioCell™ and apply sufficient culture media to cover the entire surface area. Put in place atop the stage of the inverted microscope and set the temperature control for the BioCell™ to 37°C.
5. Move the AFM head onto the inverted microscope. Using the stepper motor, lower the head into the culture medium on the glass slide (see Note 6).
6. Using the CCD-camera-equipped microscope, focus the laser on the cantilever tip. Adjust the vertical and lateral deflections in order to attain a sum value which is maximal for the cantilever.
7. Prior to approaching the sample, values for setpoint (1 V), IGain (20 Hz), and PGain (0.0001) should be inputted.
8. For force spectroscopy, an accurate determination of the cantilever spring constant is required. Firstly, the conversion factor from volt to nanometer for the deflection of the cantilever (i.e. the sensitivity) is determined. To do so, a force curve on a hard surface (e.g. a portion of the glass slide which is free of cells) is performed. The resultant force curve is fitted and a sensitivity value is determined. Next, a force spectrum is produced and a peak reflecting the cantilever resonance frequency is identified. The resonance is then fit with a Lorentz curve and used to calculate the spring constant (k). For further details, see ref. 13.
9. Using the approach settings detailed in step 7 of Subheading 3.5, approach the sample.
10. In force spectroscopy mode, an indentation test on a cell is performed. From our work, a z length of 5 µm and scan rate of 1 Hz produces a suitable force curve. Following this,

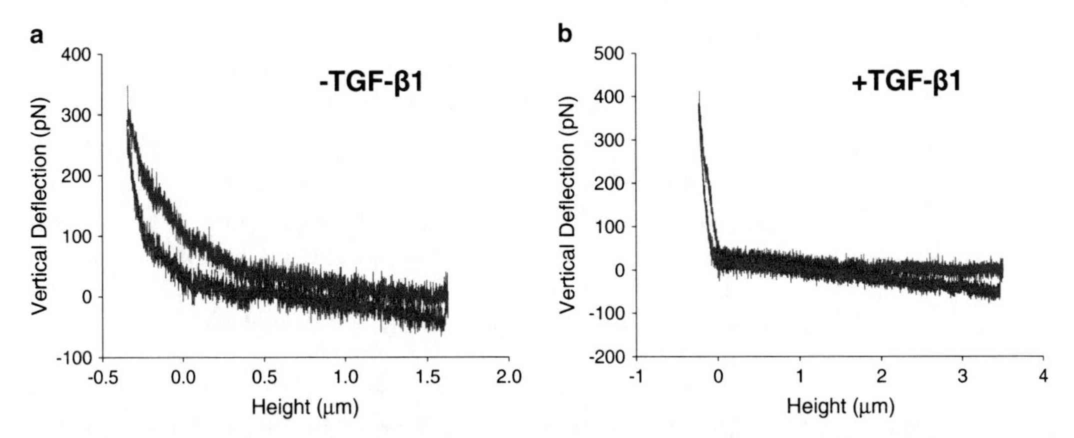

Fig. 2. Representative force versus *z* position measured in liquid for untreated and TGF-β1-stimulated A549 cells. A549 cells were grown on glass coverslips and treated with TGF-β1 (5 ng/ml) for 48 h. In contact mode, the cells were indented using a DNP tip. The force–distance curves were produced while the piezo approached the cell (*blue line*) and retracted (*red line*) at constant velocity (1.67 μm/s). In the untreated cell (**a**), the slope of the force–distance curves is relatively smoother than that corresponding to the curve of the TGF-β1-stimulated cell (**b**). This notable distinction is suggestive of a stiffer cellular structure upon exposure to TGF-β1. Due to the viscoelasticity of the cells, a hysteretic curve is generated.

force mapping mode is employed to acquire a series of 4×4 force–distance curves (at least five per cell) (see Note 7).

11. Analysis of force curve data is undertaken using JPK Image Processing software (see Fig. 2). Using the Hertz model, Young's modulus values are extrapolated from each force curve (see Note 8). During analysis, it is essential that the appropriate tip geometry (quadratic pyramid) and correct half angle to edge (22.5°) and Poisson ratio (0.3) values are chosen.

3.6. Imaging and Topographical Analysis Using Atomic Force Microscopy

Imaging using an AFM provides detailed information regarding the architecture of cells and their cytoskeletal system. By scanning across the surface of a sample, an image representing a 3D reconstruction of the sample's topographical features is produced.

1. Imaging is performed using the equipment described in steps 1 and 2 of Subheading 3.5.
2. A549 cells growing in an 8-well chamber slide are fixed as per step 1 of Subheading 3.2.
3. The slide is positioned atop the stage of the inverted microscope, and the AFM head equipped with a PPP-FM cantilever tip (NanoSensors, Neuchatel, Switzerland) is put in place. Using a pipette, carefully apply sufficient PBS to form a meniscus between the cantilever tip and the slide.
4. Focus the laser on the cantilever tip as in step 4 of Subheading 3.5.
5. As per step 8 of Subheading 3.5, determine the sensitivity of the cantilever tip.

6. Using suitable approach settings (e.g. setpoint [0.75 V], IGain [50 Hz], and PGain [0.002]), the AFM cantilever is approached to the sample until it contacts the surface.
7. Locate the desired cell(s) to be imaged using the CCD-camera-equipped microscope.
8. Select Intermittent Contact mode (fluid). Engage the cantilever tip and commence scanning. The quality of the image is dependent on selection of appropriate image settings, in particular, that of IGain, PGain, and amplitude setpoint. The IGain and PGain facilitate determination of the signal-to-noise ratio. The amplitude setpoint governs the degree of hardness with which the cantilever is touching the sample. If it is too light, the resultant image is devoid of sufficient details, while if it is too hard, the tip may damage the surface (see Note 9).
9. Once the optimum settings for the sample are found, the imaging is re-commenced to ensure artefacts of sub-optimum settings are avoided.
10. Using JPK Image Processing software, desired analysis of images is performed. Details regarding topographical features can be ascertained from the "Section" tool which produces cross-sectional profiles of selected areas within an image. The "Histogram" tool permits quantitative analysis of the degree of surface roughness, including root mean square (RMS), roughness (R_q), and peak-to-valley roughness (R_t).

4. Notes

1. For morphological analysis using HCA microscopy, it is essential that an appropriate seeding density is used. This is dependent on the cell type and the nature of the experiment and should be determined empirically. In general, relatively low seeding densities should be used as they permit cells to grow in an uncrowded environment. This ensures effective segmentation of analysed images following time in culture.
2. FBS contains numerous growth factors which can adversely affect the results of experiments. To minimise these potential effects, experiments should be performed using a low-serum containing medium (i.e. 1%).
3. When analysing cytoskeletal structure, the fixative method employed should ensure minimal structural distortion and retain all cytoskeleton-associated material. In this regard, the slow addition of paraformaldehyde (8% w/v) to cell media and incubation at 37°C yields excellent structural preservation of the actin cytoskeleton.

4. Phalloidin will dissociate over time into the PBS during storage leading to background autofluorescence which can obscure fine structural details. Thus, plates should be imaged within a few days of staining.
5. The IN Cell™ Analyzer 1000 Workstation morphology analysis module utilises a segmentation technique which divides each image into discrete objects and/or regions. Key to this process is selection of the appropriate method of segmentation. Top-hat is a quick and efficient transformation which emphasises objects of a particular size, enhancing their detection. This method is particularly useful for identification of objects which exhibit a uniformity of size and shape (e.g. nuclei). However, cell bodies show a variety of size and cannot be effectively identified and analysed by the top-hat method. Instead, the multiscale top-hat method is employed. By providing a characteristic cell body area, the software can calculate two scales of detection for top-hat transformation, and in this way permit efficient analysis. In both cases, a suitable level of sensitivity detection should be chosen. The sensitivity setting governs which pixel clusters qualify as objects as determined by their intensities relative to local background. Thus, as the sensitivity is increased, the possibility of detecting dimmer objects is enhanced.
6. When lowering the AFM head into cell culture medium bubbles of air may become trapped. This can prevent efficient operation of the AFM. If any air bubbles are detected, the AFM head should be removed and the tip holder detached. Using a Kimwipe®, the tip can be blotted dry and re-mounted.
7. The cytoskeleton exhibits considerable regional variability. As a result, single force–distance curves are insufficient for the purpose of characterising cellular mechanical properties. Instead, an array of indentations across an area of interest within the cell can provide an efficient assessment of a cell's elastic properties. When performing force measurements, both the nucleus and cell periphery should be avoided as these regions are characterised by notable softness and rigidity, respectively, and as such, are not reflective of the cell as a whole.
8. The Hertzian model used to calculate the Young's modulus (E) of a cell assumes that the sample is isotropic and linearly elastic. Moreover, it supposes that the indenter cannot be deformed and that there exists no further interactions between indenter and sample. However, in fact, cellular samples often show viscoelastic behaviour and are inhomogeneous. In particular, this model assumes that indentation is negligible when compared to the thickness of the sample. Thus, the depth of indentation must be optimised and should not exceed 5–10% of the height of the cell. Despite the aforementioned limitations to this

model, provided due consideration is taken, satisfactory and reproducible results can be generated.

9. The IGain and PGain represent the integral and proportional gains for the adjustment of the height feedback loop. Attainment of optimal gain values is dependent on the scan size and rate, selected setpoint and *z* range and the topography of the sample. Typically, higher gains ensure that the cantilever tip accurately follows the sample surface, generating a superior topography. Moreover, it obviates potential damage to the tip or sample since the tip can react speedily to any alterations in sample height.

Acknowledgements

STB is funded by an IRCSET Government of Ireland Postgraduate Scholarship in Science, Engineering and Technology. This work has been funded in part by a Strategic Research Cluster grant (07/SRC/B1154) under the National Development Plan co-funded by EU Structural Funds and Science Foundation Ireland.

References

1. Lee, J.M., Dedhar, S., Kalluri, R., Thompson, E.W. (2006) The epithelial-mesenchymal transition: new insights in signaling, development, and disease. *J Cell Biol.* **172**, 973–81.
2. Kalluri, R., Neilson, E.G. (2003) Epithelial-mesenchymal transition and its implications for fibrosis. *J Clin Invest.* **112**, 1776–84.
3. Guarino, M., Rubino, B., Ballabio, G. (2007) The role of epithelial-mesenchymal transition in cancer pathology. *Pathology.* **39**, 305–18.
4. Radisky, D.C., Kenny, P.A., Bissell, M.J. (2007) Fibrosis and cancer: do myofibroblasts come also from epithelial cells via EMT? *J Cell Biochem.* **101**, 830–9.
5. Kalluri, R., Weinberg, R.A. (2009) The basics of epithelial-mesenchymal transition. *J Clin Invest.* **119**, 1420–8.
6. Savagner, P. (2001) Leaving the neighborhood: molecular mechanisms involved during epithelial-mesenchymal transition. *Bioessays.* **23**, 912–23.
7. Wendt, M.K., Allington, T.M., Schiemann, W.P. (2009) Mechanisms of the epithelial-mesenchymal transition by TGF-beta. *Future Oncol.* **5**, 1145–68.
8. Thoelking, G., Reiss, B., Wegener, J., Oberleithner, H., Pavenstaedt, H., Riethmuller, C. (2010) Nanotopography follows force in TGF-beta1 stimulated epithelium. *Nanotechnology.* **21**, 265102.
9. Shahin, V., Barrera, N.P. (2008) Providing unique insight into cell biology via atomic force microscopy. *Int Rev Cytol.* **265**, 227–52.
10. Rausch, O. (2006) High content cellular screening. *Curr Opin Chem Biol.* **10**, 316–20.
11. Lieber, M., Smith, B., Szakal, A., Nelson-Rees, W., Todaro, G. (1976) A continuous tumor-cell line from a human lung carcinoma with properties of type II alveolar epithelial cells. *Int J Cancer.* **17**, 62–70.
12. Small, J., Rottner, K., Hahne, P., Anderson, K.I. (1999) Visualising the actin cytoskeleton. *Microsc Res Tech.* **47**, 3–17.
13. Hutter, J.L., Bechhoefer, J. (1993) Calibration of atomic-force microscope tips. *Rev Sci Instrum.* **64**, 1868–73.

INDEX

A

Analysis
- construction 139–152
- map 143, 144
- sectioning 147, 151, 156
- slide staining 147

Antibody 57, 63, 72, 110, 112, 114, 115, 117, 118, 120, 141, 148, 155–159, 162–166, 171, 173–175, 178, 179, 184, 185, 191, 192

Antigen retrieval 161, 165

Archived samples 78, 90, 94

B

BeadArray 77–79, 82, 89

Beta cells
- function 16

Biomarker 1, 79, 94, 123, 147, 148

Bone marrow 56, 67, 72, 172, 174

C

Cancer 1, 3, 9, 15, 35, 37, 47, 55–73, 78, 90, 99, 110, 124, 127, 141, 143, 182, 198

Cell culture 4, 5, 17, 110–111, 170–171, 182, 183, 199, 200, 207

Cell line 3, 17, 22, 31, 58, 71, 94, 124, 127, 128, 134, 139–152, 179, 182, 186, 189, 192–194, 200

Cell line array (CMA) 139–152

Circulating tumour cells (CTCs) 55–73

Clustering
- hierarchical 36–38, 43–46, 97, 148
- *k*-means 38, 45–47

CMA/TMA 139–152

Complementary DNA (cDNA) 4, 6–8, 10, 16, 19, 20, 22–24, 27, 31, 32, 34, 44, 46, 78, 79, 83, 84, 91, 92, 94, 96, 101–104

Conditioned medium (CM) 16–20, 22–25, 182, 183, 186–188, 192–194

Cytoskeleton 164, 198, 200, 206, 207

D

DASL (cDNA-mediated annealing, selection, extension and ligation) assay 77–97

Data analysis 27–39, 41–52, 90, 91, 203

Densitometry 21–23, 114, 115, 118–119

Deoxynucleoside triphosphate (dNTP) 4, 6, 8, 18, 20, 101–103, 105

Deoxyribonucleic acid (DNA)
- DNA chip (dCHIP) 33, 43–44

Differentially-expressed (DE) 32–35, 39, 44–46, 49, 50, 92, 94, 124, 133

E

Electrospray ionization (ESI-MS/MS) 124

Epithelial-mesenchymal transition (EMT) 197–208

Ethidium bromide 4, 6, 18, 21

Exosomes 181–194

Experiment
- double channel 29–30
- single channel 30–31

Experimental design 27–39, 45, 91, 202

F

Fixatives
- acetone 166, 171, 178
- methanol 166, 171, 178
- methanol-acetone 178

Fluorescent resonance energy transfer (FRET) 3

Formalin-fixed paraffin-embedded (FFPE) 58, 71, 77–97

G

Gel electrophoresis 4, 7, 10, 21, 111, 112, 123–136, 182

Gene expression analysis 1, 11, 27, 35, 41, 43–48, 51, 55–73, 77–97, 99, 139–152, 170, 182

Gene ontology (GO) 39, 43, 44, 48, 49

H

HEK293T cells 3, 6, 8, 9

High-content analysis (HCA) 197–208

Lorraine O'Driscoll (ed.), *Gene Expression Profiling: Methods and Protocols*, Methods in Molecular Biology, vol. 784, DOI 10.1007/978-1-61779-289-2, © Springer Science+Business Media, LLC 2011

I

Image acquisition system 42, 118, 191, 201–202
Image analysis 21, 32, 35, 42, 118, 120, 140, 148, 149, 176, 201
Immunofluorescence 155–156, 172, 179, 199, 200
Immunohistochemistry (IHC), 58, 71, 140, 141, 144, 147–149, 155–166, 179
Insulin-producing cells 16

M

Mass spectrometry 123–136, 182
Matrix assisted laser desorption ionization time-of-flight mass spectrometry (MALDI-ToF MS) 124
Microarray
- microarray data analysis system (MIDAS) 44–45
- microarray data manager (MADAM) 44
- significance analysis of microarrays (SAM) 45–46

microRNA (miRNA) 1, 99–107, 181–194
Microscopy
- advanced microscopy 169–180
- atomic force microscopy (AFM) 197–208
- fluorescence microscopy 160, 169
- laser scanning confocal microscopy (LSCM) 169–180
- transmission electron microscopy (TEM) 185–186, 189, 192–193

MIN6
- MIN6 (H) 16
- MIN6 (L) 16
- MIN6 B1 16, 17, 23

Moloney murine leukemia virus (MMLV) 18, 20, 22
mRNA 1, 3, 7, 9, 10, 15–25, 27–39, 78, 79, 99, 181–194
Multivesicular bodies (MVBs) 182

N

NanoDrop 18, 24, 103
NIH-3T3 110, 113
Normalisation 2, 32–33, 42–47, 100
Nucleic acids
- extracellular 15, 24

P

Pathway analysis 39, 44
Peptide mass fingerprints (PMF) 124
Peripheral blood mononuclear cells 68, 69
Plasma 15, 24, 69, 171, 177, 179, 182, 187
Polymerase chain reaction (PCR)
- Black Hole Quencher (BHQ) 2, 5, 8
- carboxy-X-rhodamine (ROX) 8
- multiplex 1–11, 182
- primers and probes design 4
- real-time/quantitative 1–11, 102, 104–107
- reverse transcriptase (RT) 1, 3, 15–25, 182
- SybrGreen 2
- *Taq* DNA polymerase enzyme 18, 100
- TaqMan probes 2, 3, 100
- threshold cycle (VIC) 2, 8, 11, 100

Principal component analysis (PCA) 38–39, 45, 46
Protein expression profiling 140
Proteomics 123, 124, 136, 182

Q

Quality control (QC) 4, 42–44, 78, 140, 148, 150

R

Receptor tyrosine kinase (RTKs) 109–120
Replicates 25, 30, 31, 34, 91–93, 95
Retro-orbital bleed 58, 68–69
Ribonucleic acid (RNA) 1, 4–10, 15–20, 22–25, 28, 29, 31, 32, 58, 67, 69, 70, 73, 78, 79, 81, 82, 83, 90–97, 100–107, 182
- isolation
  - cells 5
  - tissues 6

RNAse-free
- DNase 10, 17, 20
- water 6, 17, 19, 20, 88, 101, 103, 106

S

Saliva 24, 124, 127, 182, 187
SDS-polyacrylamide gel electrophoresis (SDS-PAGE), 110–112, 116, 119, 125, 126, 131, 132, 136, 184, 185
Serum 4, 15–17, 24, 56, 106, 110, 112, 113, 119, 124, 127, 128, 162, 163, 165, 171–173, 175, 182–184, 186–189, 192–194, 199, 200, 206
Software 4, 8, 18, 21, 28, 32, 41–52, 89, 96, 118, 120, 124, 126, 133, 135, 143, 176, 201, 202, 205–207
Stem cell 55, 56, 63, 67
Surface markers 56, 65, 66

T

TaqMan low density microRNA array (TLDA) 101, 102, 105–107
Thermal cycler 2, 4, 8, 16, 18, 85
Tissue arrayer 145, 150, 151
Tissue microarray (TMA) 139–152
Tumour initiating cell 55, 56, 60, 61, 63, 71

Two-dimensional difference gel electrophoresis (2D DIGE)...........124, 128, 129, 131, 133, 134, 136

U

Urine...182, 186

W

Western blotting...109–120

X

Xenograft model...61